T0186304

Clinical Transfusion Medicine

Joseph D. Sweeney, M.D.
The Miriam and Roger Williams Hospitals
Brown University School of Medicine
Providence, Rhode Island

Yvonne Rizk, M.D.
Women and Infants Hospital
Providence, Rhode Island

LANDES
BIOSCIENCE
AUSTIN, TEXAS
U.S.A.

VADEMECUM
Clinical Transfusion Medicine
LANDES BIOSCIENCE
Austin

Copyright © 1999 Landes Bioscience

Please address all inquiries to the Publisher:
Landes Bioscience, 810 S. Church Street, Georgetown, Texas, U.S.A. 78626
Phone: 512/ 863 7762; FAX: 512/ 863 0081

ISBN: 1-57059-494-5

Library of Congress Cataloging-in-Publication Data

Sweeney, Joseph, 1952-
    Clinical transfusion medicine / Joseph D. Sweeney, Yvonne Rizk.
      p. cm.
    "Vademecum."
    Includes bibliographical references and index.
    ISBN 1-57059-494-5
    1. Blood--Transfusion handbooks, manuals, etc. I. Rizk, Yvonne. II. Title.
    [DNLM: 1. Blood Transfusion Handbooks. WB 39 S974c 1999]
RM171.S926 1999
615'.39--dc21
DNLM/DLC   99-24311
for Library of Congress   CIP

Contents

Preface

Clinical transfusion medicine is an evolving subspecialty, which straddles traditional areas of pathology and clinical hematology. This subspecialities is concerned with aspects of blood procurement, including safety, logistics and economics, and the appropriate use of blood products in different clinical situations. This causes the transfusion medicine physician to interact with (and occasionally come into conflict with!) surgeons, anesthesiologists, internists, and many subspecialists in internal medicine, particularly, oncologists and hematologists. The resultant of this interaction should be improvement in blood utilization. In short, the role of clinical transfusion medicine is to promote good transfusion practice.

Promoting good transfusion practice is often hindered by a lack of good clinical data validating many current transfusion practices. Confounding this problem is the often entrenched belief in the clinical usefulness of many traditional transfusion practices. The transfusion medicine physician is, therefore, frequently put in the position of altering practices, a precarious role in any institution!!

This short book is intended to put clinical problems in perspective as they relate to decision making regarding blood transfusion. It is aimed at nursing staff, perfusionists, nurse practitioners, physician assistants and medical students and residents, many of whom lack depth in their understanding of transfusion. The language is kept as nontechnical as possible therefore, and detail is intentionally omitted. However, it is hoped that the background information and general principles should facilitate the exercise of good judgment.

Joseph D. Sweeney, M.D.
Yvonne Rzik, M.D.
Providence, Rhode Island
May, 1999

Acknowledgments

The expert assistance of Ms. Susan Sullivan in the preparation of this monograph is gratefully acknowledged.

Introduction

The scope of transfusion medicine can be separated into two definable areas of activity (Fig. 1.1). First, there are those activities concerned with the *manufacture of blood products*. These processes occur mostly in Community Blood Centers or Fractionation plants. The 'source material' is obtained from healthy human subjects, known as blood donors. This part of transfusion medicine is concerned with the collection, processing, and testing of blood donations and the maintenance of an inventory of blood products prior to shipping to sites of transfusion. The kinds of activities are similar to those which occur in standard pharmaceutical houses. Emphasis is on the potency, safety, efficacy, and purity of the manufactured blood products.

The second area of transfusion medicine can be described as *clinical transfusion medicine*. Clinical transfusion medicine is concerned with aspects related to the transfusion of blood products to recipients. The human subjects of interest are sick patients, and called blood transfusion recipients. Emphasis is on product availability, appropriateness of use, informed consent, compatibility testing, administration of blood, monitoring for adverse events (called transfusion reactions) and the long term follow up for complications of infectious disease. These differences are shown in Table 1.1 but are really a continuum, as illustrated in Figure 1.1.

This book is concerned with the second area of transfusion medicine i.e., clinical transfusion medicine. Brief reference will be made to manufacture, however, where background information is important. Clinical transfusion medicine mostly occurs in a hospital setting, although other sites of transfusion are becoming commonplace, such as outpatient departments, renal dialysis units, physician offices or even the recipient's home. Within the hospital structure, the focal area for this activity is the blood bank. Although blood banks are concerned with the dispensing of a therapeutic product, they are often part of pathology laboratories. From a theoretical perspective it would be more appropriate if blood banks were more closely linked to hospital pharmacies. A comparison of pathology laboratories, blood banks, and pharmacies in Table 1.2 illustrates this point. The historical reason for blood banks to be within departments of pathology, and not part of a pharmacy, primarily relates to the need to perform compatibility testing as this testing is similar to other kinds of tests traditionally performed by laboratory technologists.

The purpose of this book is to serve as a quick source of useful, practical information for the many aspects of clinical transfusion medicine. The content reflects practice in the United States, but is generally applicable to other countries. Knowledge of transfusion medicine is surprisingly limited even among experienced hematologists and pathologists and a simple rapidly readable text serves a useful

Clinical Transfusion Medicine, by Joseph D. Sweeney and Yvonne Rizk. © 1999 Landes Bioscience

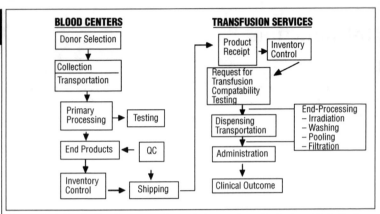

Fig. 1.1. Activities in blood centers and transfusion services. Product quality is the focus of blood centers; clinical outcome is the focus of transfusion services.

Table 1.1. Comparison of the major areas of activities in transfusion medicine

	Manufacture of blood products	Clinical transfusion medicine
Site	Blood center or plasma fractionation plant	Blood bank/ transfusion service
Activity	Manufacture of blood components and derivatives	Transfusion of blood components and derivatives
Regulatory agencies/ accrediting (USA) organizations	FDA AABB CAP	JCAHO AABB
Product specifications	Defined	General terms
Variation in practices between sites	Minimal	Large variations in techniques/and practices
Auditing of practices	Standard performed	Variable; often not
Human population	Healthy subjects (blood donors)	Ill Subjects (blood recipients)
Product focus	Potency; efficacy; safety; purity;	Availability; appropriateness of use and administration effectiveness adverse events

FDA = Food and Drug Administration
AABB = American Association of Blood banks
JCAHO = Joint Commission on Hospital Accreditation
CAP = College of American Pathologists

Table 1.2. Comparison of diagnostic pathology laboratories, blood banks and pharmacies

	Pathology Laboratories	Blood banks	Pharmacies
Product	Diagnostic	Therapeutic	Therapeutic
Personnel	Technologists/	Technologists/ Technicians	Pharmacists Technicians
Specialist Physicians	Pathologists	Pathologists/ Hematologists	Clinical Pharmacologists
Regulatory/ Accrediting Agencies	CAP; JCAHO	CAP; AABB; JCAHO; FDA	JCAHO; FDA

CAP = College of American Pathologists; AABB = American Association of Blood banks JCAHO = Joint Commission on Hospital Accreditation; FDA = Food and Drug Administration

need. It is divided into sections, each consisting of short chapters which can be read within five minutes addressing specific clinical situations, and facilitation of rapid clinical decision-making or giving essential background information are the objectives of the book. It is not intended as a reference text in transfusion medicine, therefore, and the readership is aimed at medical students, residents in training, or nursing and allied health personnel. Chapters are, therefore, intentionally short without specific references. When more detailed information on a specific clinical situation is required, it is suggested that an electronic search be conducted or reference textbooks such as may be conveniently available in the Blood bank should be consulted. Suggested sources of information with more detail are given in the Appendix.

Allogeneic Blood Products

The term *allogeneic* refers to blood products manufactured from blood donations from healthy subjects (blood donors) which are intended for transfusion to different subjects (blood recipients). In the past these products were called "homologous", but the current preferred term is allogeneic, in order to be consistent with solid organ transplantation terminology. Other names used regionally to describe these products are; "regular blood", "shelf blood" or "banked blood".

The term Blood Product, then, is an all embracing term used to describe any end-product produced from human blood. First, there is *whole blood*, which is collected into a solution that functions both to anticoagulate and preserve the red blood cells. These are simple solutions containing citric acid to chelate calcium and, therefore, prevent activation of the coagulation system and glucose to allow red cells to metabolize during in vitro storage (e.g., citrate-phosphate-dextrose or CPD). Adenine may also be present (CPD-A1), which improves red blood cell adenosine triphosphate (ATP) levels. Anticoagulated whole blood is collected into, stored in, and transfused from, its primary container. Although whole blood was commonly used prior to the 1970s, this has diminished over the past decades. The two remaining clinical situations where whole blood is still preferentially requested for transfusion are patients with trauma requiring multiple transfusions and cardiac surgery, particularly pediatric cardiac surgery. This is because it has been suggested that 'fresh' whole blood (less than 48 hours old) may be preferable to correct any coagulopathy which may develop in these patients . However, the practical logistics of having fresh whole blood routinely available makes this difficult to achieve in practice.

Second, most whole blood donations are further processed by centrifugation into a number of *blood components*. Each whole blood donation is capable of producing up to five different components, but commonly, either red blood cells and plasma, *or* red cells, plasma and platelets are produced. Further processing of a unit of plasma can produce a unit of cryoprecipitate and a cryosupernatent, the latter known as cryo poor plasma. Some fibrin glue preparations are similar to cryoprecipitate, except that the process may be modified to enhance fibrinogen yields (see Chapter 23). In practice, almost all allogeneic whole blood donations are processed into at least two components, such as a red cell concentrate and plasma. The red cell concentrate can be stored in the anticoagulant plasma alone (CPD red blood cells; CPD-A1 red blood cells), or a crystalline solution can be added, which contains glucose and stabilizers to maintain the quality of red cells during the storage period (Adsol®, Nutricell®, or Optisol®). The maximum duration of storage for red cells (at 1-6°C) under such circumstances is currently 42 days (Chapter 27). Much of the plasma produced from the whole blood dona-

Clinical Transfusion Medicine, by Joseph D. Sweeney and Yvonne Rizk. © 1999 Landes Bioscience

tions is shipped to fractionation plants for further manufacture into blood derivatives. Some of the plasma, however, is retained in blood centers in the frozen state and used for clinical transfusion purposes (see Chapter 29).

A single unit of platelets may also be produced from each blood donation. Although the terminology is confusing, a platelet unit derived from a whole blood donation is commonly known as a random donor platelet. A separate type of blood component is an apheresis blood component. Several different types of apheresis components are available but the most important is the platelet apheresis product, commonly known as single donor platelets. As indicated above, a unit of random donor platelets is also derived from a "single donation" and hence the terminology is confusing (Chapter 28). Other allogeneic apheresis products may be more widely available in the future, as a 'double unit' of red cells, or combinations of platelets and red cells, or red cells and plasma, may be obtained from a single donor using these devices. It is anticipated that many such products will be approved for use by late 1999. The hallmark of a blood component is that is derived from a single donation. Each such donation has a unique identification (unit number or lot number).

Third, there are blood products known as *blood derivatives*. These are manufactured from a plasma pool usually containing between 5,000 and 20,000 donations. This plasma pool constitutes a new lot number, composed of the individual lot numbers of each donation which makes up the pool. All blood derivatives in current use are acellular products. Derivatives in common use are 5% or 25% albumin, immunoglobulins, and the plasma derived coagulation factor concentrates. Since blood derivatives are produced from such a large number of blood donations, there is always the ongoing concern that new viruses from apparently health donor(s), may enter each pool and potentially infect a large number of recipients. This was responsible for the spread of hepatitis in the 1970s and, subsequently, human immunodeficiency virus (HIV-1) in the 1980s in the hemophilic population. Blood derivatives are routinely subjected to a variety of processing steps. Some of these steps are intentionally performed to destroy viruses, which is called *viral attenuation*. Use of at least two different types of viral attenuation processes is now common in order to optimize the destruction of viruses. Examples of these processes are pasteurization i.e., heating to 60°C for ten hours; various separation steps, e.g., gel filtration or micro-filtration, and chemical treatments such as solvent detergent exposure. In spite of the clear effectiveness of these viral attenuation processes, there is still the potential for some viruses to be resistant to these steps and result in the infection of blood transfusion recipients.

The last type of allogeneic blood product is stem cell products. Stem cells can now be collected from a number of sources, other than the traditional bone marrow, such as peripheral blood or umbilical cord blood. Allogeneic stem cell products are always derived from a single donor, but multiple donations may be required if peripheral blood is the source. Since stem cells are generally transfused in specialized transplant units, they will not be described further and the reader is referred to the *Clinical Handbook of Bone Marrow Transplantation* for further information on these products and associated technology. These products are illustrated in Figure 2.1.

The physical state of blood products during storage varies with the product type. Red cells and platelets are typically stored in the liquid state. Plasma and cryoprecipitate are stored in the frozen state. Certain blood products are manufactured and stored in the lyophilized state; examples are some immunoglobulin preparations and coagulation factor concentrates. However, red cells (or less commonly platelets) may also be stored in the frozen state using cryoprotective agents; and certain blood derivatives, such as one preparation of immunoglobulin and all albumin preparations, are stored in the liquid state. All products are transfused as a liquid product, either after thawing of frozen products or reconstitution of lyophilized products. This accounts for the delay in availability since it takes between 15-30 minutes to either thaw frozen products, or to reconstitute lyophilized products prior to transfusion. This is shown in Table 2.1.

The allogeneic blood supply in the United States, much of Europe and Japan is predominantly collected from what are known as 'volunteer blood donors'. These donors donate for altruistic reasons and do not receive any remuneration or reward of a material monetary value. Platelet pheresis donors in some centers receive a token remuneration. However, much of the plasma collected for fractionation comes from paid donors. A different type of volunteer blood donor is known as a directed donor, and the donation as a 'directed donation'. Directed donor blood products are allogeneic blood products which meet all the requirements for

Source:	Healthy Humans	
	Donors	
	Donation (= individual lot #)	
Products:	Whole blood	
	(single donation)	
	Blood components	Red Blood Cells
	(single donation)	-plasma → cryoprecipitate; cryo poor plasma
		-platelets
		-apheresis components
	Blood Derivatives	Albumin
	(5,000-20,000	Immunoglobulins
	donations)	Coagulation Factors
	Stem Cells	Bone Marrow
	(single/multiple	Peripheral Blood
	donations)	Umbilical Cord
		Fetal Hepatocytes
Physical State	Liquid (e.g., red cells, platelets)	
	Frozen (e.g., plasma)	
	Lyophilized (e.g., derivatives)	

Figure 2.1. Allogeneic blood products

Table 2.1. Some properties of allogeneic blood products in common use

Product	Physical State	Approx. Volume (mls)	Storage Temperature	Shelf Life	Comments
Whole blood in CPD	Liquid	525	1-6°C	21 days	Hematocrit approx. 35
Whole blood in CPD-A1	Liquid	525	1-6°C	35 days	Hematocrit approx. 35
Red Blood Cells in CPD	Liquid	300	1-6°C	21 days	Hematocrit < 70
Red Blood Cells in CPD-A1	Liquid	300	1-6°C	35 days	Hematocrit < 70
Red Blood Cells in Preservative	Liquid	350	1-6°C	42 days	Hematocrit 50-60; little plasma
Random Donor Platelets (RDP)	Liquid	50	20-24°C	5 days	4-10 units pooled into a single container
Single Donor Platelets (SDP)	Liquid	180-350	20-24°C	5 days	Equivalent to 6-8 units of RDP
Fresh Frozen Plasma	Frozen	220	-18°C or lower	1 year	15-30 minutes to thaw
Cryoprecipitate	Frozen	5-15	-18°C or lower	1 year	Thawed, then pooled
Vial of Coagulation Factor	Lyophilized	10-20	Refrigeration	1 year	Reconstituted with diluent, then transfused

CPD = Citrate – Phosphate – Dextrose
CPD-A1 = Citrate – Phosphate – Dextrose – Adenine

a standard allogeneic blood product. However, they differ in that the intended recipient is identified at the time of donation. The practice of directed blood donation is often performed in the context of donating blood for a relative or friend in anticipation of surgery or cancer chemotherapy. Directed donations are generally neither encouraged nor discouraged by blood collection facilities. There is no evidence that they are any more safe (i.e., less likely to transmit viral infections) than the non-directed volunteer blood supply. On the contrary, since many directed donors are first time donors, there is a higher prevalence of viral disease markers, raising concern regarding a possible increased risk. Directed donor blood may be transfused to a recipient other than the intended recipient, if the latter does not require transfusion, a practice *"called crossover"*.

All allogeneic blood donations are routinely tested for syphilis and viral disease markers as shown in Table 2.2.

Table 2.2. Testing of blood donations for microbial diseases

Test	Year Initiated
Serological Test for Syphilis	1949
Hepatitis B Surface Antigen (HBs As)	1972
Antibody to HIV-1	1985
Antibody to Core Antigen of Hepatitis B (Anti HBc)	1986
* Alanine Aminotransferase (ALT)	1986
Antibody to HCV	
Generation I	1990
Generation II	1992
P24 Antigen of HIV-1	1996
Nucleic Acid Testing for HCV, and HIV-1	1999

* No longer routinely required

Autologous Blood Products

Autologous blood products differ from allogeneic products in several important respects. First, the source of the product may not always be a healthy human donor but rather, a patient with an anticipated need for blood products in the near or immediate future. Second, criteria for accepting blood donations in most collection sites differ between allogeneic and autologous blood donors, with more liberal criteria being applied to autologous donors. Third, the autologous blood donation is a special type of directed blood donation in that the donor is the intended recipient and unlike directed donor units, it is an uncommon practice to use autologous units for transfusion to a recipient other than the intended recipient (crossover, Chapter 2). Fourth, autologous products differ substantially from allogeneic products in composition, potency and shelf life. Different types of autologous products and some characteristic features are shown in Figure 3.1.

PREDEPOSIT AUTOLOGOUS BLOOD (PAD)

This type of blood product most closely resembles the standard allogeneic whole blood donation. The blood may be collected and retained as a unit of unprocessed whole blood, but it is much more common to process the donation into a unit of red blood cells and plasma. The red cells are often stored in an additive solution, which extends the shelf life to 42 days (see Chapter 2). The disposition of the plasma varies. It may be made available for use by the autologous donor. It is also possible to ship this plasma to fractionation plants, if the autologous blood donor meets all the standard criteria for allogeneic blood donation. Occasionally, the autologous plasma can be used to manufacture cryoprecipitate or a fibrin glue concentrate for intraoperative use, for example, in vascular or cardiac surgery.

Predeposit autologous blood (PAD) is donated within the six week period prior to intended use, but most commonly within 3-4 weeks of surgery. It is the most common form of autologous blood product. Units are collected generally at weekly intervals, but at not less than three day intervals, and not within 72 hours of the intended time of surgery. These products are tested for standard infectious disease markers, as in the case of allogeneic units (Chapter 2). An important difference, however, is that the presence of a positive infectious disease marker tests (which always precludes the shipping of an allogeneic blood product), may allow the shipping of the autologous units for transfusion. Such units will have a biohazard label attached. Suitable patients for PAD are shown in Table 3.1.

Rarely, platelets are donated in a predeposit context, using apheresis devices, and under these circumstances, the platelets are cryopreserved. The only practical use of this *uncommon practice* is in the management of patients in remission of

Clinical Transfusion Medicine, by Joseph D. Sweeney and Yvonne Rizk. © 1999 Landes Bioscience

Procedures and Types of Products	Component	Volume (mls)	Shelf Life
(a) Predeposit Autologous Donation: (PAD)	Whole blood	525	35 days
	Red Cell Component Sometimes Plasma/ Cryoprecipitate or (fibrin glue) cryo- preserved platelets	350	41 days
(b) Preoperative Hemodilution: or	Whole blood	475-565	8 hours
Preoperative Apheresis	Platelets Plasma	180-400	8 hours
(c) Intraoperative Salvage (Processed or Unprocessed)	Red Blood Cells	variable	8 hours
(d) Post Operative Salvage	Red Blood Cells	variable	?
(e) Stem Cell Products	Bone Marrow Peripheral Blood		
Physical State: Liquid (Red Cells/Platelets) Frozen (Plasma/Cryopreserved Platelets)			

Fig. 3.1. Autologous blood products; Source: patients requiring surgical/medical treatment

---

Table 3.1. Patients suitable and unsuitable for predeposit autologous donation

---

I. Elective
   1. Orthopedic or urologic surgery
   2. Elective vascular or cardiac surgery
   3. 'Elective' abdominal procedures (e.g. colorectal surgery)

II. Performed primarily to allay anxiety, but unlikely to be of value
   1. Obstetrical patients
   2. Prior to minor cosmetic procedures or minor surgery (e.g. lumpectomy)

---

acute myelogenous leukemia, in anticipation of use during consolidation therapy or subsequent bone marrow transplantation.

PREOPERATIVE HEMODILUTION OR PREOPERATIVE APHERESIS

Preoperative hemodilution or preoperative apheresis is essentially the same type of procedure. Anticoagulated blood is collected immediately before (i.e., within 2 hours) a surgical procedure. The end product of preoperative hemodilution is a unit of unprocessed whole blood in CPD as anticoagulant. Preoperative apheresis is a procedure in which blood components, most commonly platelets, but sometimes plasma, are collected preoperatively with the intention of transfusion, usually towards the end of the surgical procedure. Examples of components collected in this category are platelets collected prior to cardiac surgery with the intent of reinfusion immediately subsequent to protamine neutralization; or plasma collected preoperatively and reinfused in the same situation. These apheresis autologous products have also been used in orthopedic and vascular surgery, although this is not as well studied. In addition, although the intraoperative time period is short, plasma collected preoperatively can be processed further in the operating room into a fibrin glue preparation, using special devices, which are capable of rapid freezing and thawing. Preoperative hemodilution, and particularly preoperative apheresis, are not standard in many surgical centers. The types of patients who are candidates for this procedure are similar to those who predeposit autologous blood, i.e., patients undergoing elective orthopedic, urologic or vascular surgery, and cardiac patients who are unsuitable for predeposited autologous donation (Table 3.1).

INTRAOPERATIVE SALVAGE

An autologous blood product can also be produced from red blood cells which are salvaged (i.e., collected) intraoperatively from a site of surgical bleeding site. Red blood cells, which are shed from the surgical field, are anticoagulated and collected into a reservoir. Anticoagulation is achieved by adding heparin to the reservoir or by the addition of citric acid concurrent with aspiration. When the blood in the reservoir achieves a critical volume (600 to 800 ml), the aspirated blood can be returned to the patient, either as unprocessed blood using a filter or as processed red cells, usually using a washing technique. Unprocessed salvaged blood collected is simpler, but there is always concern with regard to contaminating cellular debris, particulate matter or activated clotting factors. Washing devices operate on the principle of centrifugation and the end product is composed of autologous red cells suspended in saline, with a volume of approximately 250 ml and an Hct of 40-60. These red cells are returned to the patient intraoperatively, if possible, or in the immediate postoperative period. All such returned blood is routinely filtered to remove white cell clumps and large particulate matter. If transfused outside of the operating room, proper identification of the unit and a written expiry time is critical. Patient populations suitable for this procedure are shown in Table 3.2.

POSTOPERATIVE SALVAGED BLOOD

A different type of blood product is that derived from postoperative salvage. This is most often collected from drainage sites after orthopedic surgery or from

---

Table 3.2. Patients suitable for intraoperative red cell salvage

1. Trauma (Chapter 14)

2. Cardiac surgery (Chapter 10)

3. Orthopedic and some urologic surgery (Chapter 11)

4. Vascular surgery (especially, Aorta Abdominal Aneurysectomy)

5. Liver transplantation (Chapter 12)

6. Cancer surgery (Chapter 13)

7. Miscellaneous: Rare blood types; patients with multiple red cell alloantibodies; Jehovah's witnesses

---

the chest tube in the postoperative cardiac patient. It is the type of autologous blood product *least* associated with demonstrated clinical benefit. Blood collected from these surgical sites is most often transfused using a filter but is otherwise unprocessed. Empiric clinical experience, however, seems to indicate that this is a safe practice.

AUTOLOGOUS STEM CELLS

The last important type of autologous blood product is autologous stem cells. Autologous stem cells are collected either from bone marrow or, increasingly, from peripheral blood. These products are invariably cryopreserved, (unlike allogeneic stem cell products) and therefore, thawed and reinfused at the time of transplantation. As in the case of the allogeneic stem cell products, these products are used in specialized units and will not be discussed (for further information the reader's referred to a *Clinical Handbook of Stem Cell Transplantation*).

When predeposited autologous blood became popular in the early 1980s, a practice evolved that only "healthier" donors were drawn, often patients requiring elective orthopedic surgical procedures. Many such donors were already long term allogeneic blood donors, and, if the blood was not required by the donor, it was common to use the blood for other recipients, a practice that is known as "crossover". In the late 1980s and early 1990s, the patient population from whom predeposited autologous blood was drawn broadened such that many did not fully meet standard criteria for allogeneic blood donors, and, thus, were ineligible for crossover. Because of current regulatory issues and the potential for donor misidentification, it is now a very uncommon practice to crossover autologous blood into the allogeneic blood supply.

Table 3.3. Risks associated with the transfusion of predeposit autologous blood

Complication	Risk Relative to Allogeneic Blood
1. Misidentification	Similar
2. Bacterial contamination	Higher (maybe x 5)
3. Nonhemolytic febrile reaction	Lower, but does occur
4. Hemolysis	Lower, but has been reported with hereditary red cell disorders or can occur with incorrect administration, e.g., overheating in a blood warmer.

There is a widespread misconception that autologous blood is "safe" and not associated with reactions. As shown in Table 3.3, severe and fatal reactions may occur when autologous blood is transfused, emphasizing that the decision to transfuse (either autologous or allogeneic) needs careful consideration of the clinical indication (Chapter 26).

Epidemiology of Blood Transfusion

Although this handbook is primarily concerned with the transfusion of blood components to specific individuals, it is, however, useful to appreciate the overall statistics in relation to the collection and transfusion of blood products within the United States.

Both the collection rate and red blood cell transfusion rates increased considerably throughout the 1970s and early 1980s, with a peak in 1986. Between 1986 and 1992, there was a decrease in total collections, although a pattern of a slight increase is evident in the later 1990s. The growth in blood collections during the 1970s and 1980s was related to increased demands for blood, particularly red blood cells, in the support of cardiac surgery and trauma, and platelets in the support of cancer chemotherapy and bone marrow transplantation. The plateau in the mid-1980s was driven by public concern regarding the potential for blood to transmit human immunodeficiency virus (HIV-1), which resulted in a more conservative approach to blood transfusion.

Currently in the United States, about 14 million units of whole blood are collected annually, from 8 million blood donors (approximately 3% of the population). Of these collections, approximately 12 million are processed into components for transfusion with testing losses and outdates accounting for the bulk of the two million donations untransfused. The 12 million units of whole blood donations are manufactured into approximately 27 million blood components; predominately red blood cells and plasma, and to a lesser extent, platelets. Much of the plasma manufactured is shipped to fractionation plants resulting in about 20 million blood components actually transfused in the United States on an annual basis.

Overall, about 4 million patients receive blood in the United States annually or slightly more than 1% of the population. The average number of red cell units transfused is approximately 3 per patient. Transfusion rates are higher for those in the older age quantiles and approximately 50% of the blood transfusion recipients are over the age of 65.

The red blood cell transfusion rates per thousand populations are shown in Table 4.1. Rates in the developed countries are largely similar (Western Europe, Japan, South Korea), but rates in the developing nations are significantly lower (most of Asia, Africa—3 per 1,000/year). Regional differences are evident in the U.S. The explanations for this are not entirely clear. Although population demographics and referral center locations may partially explain this, it is considered to relate to what are known as transfusion styles. Transfusion styles are practices based on preconceived ideas, often within institutions, with regard to the use of blood products in defined clinical situations. Transfusion styles result in waste and inappropriate use of blood (Chapter 9).

Clinical Transfusion Medicine, by Joseph D. Sweeney and Yvonne Rizk. © 1999 Landes Bioscience

Recipient subpopulations are shown in Figure 4.1. Overall, about 50% of red cells are transfused in Surgery, the majority of units being transfused in Cardiac, Orthopedic, Vascular and Urologic Surgery and Trauma patients. In Medicine, cancer treatment, bone marrow transplantation, and gastrointestinal bleeding account for most of the red cells transfused. For other products, good data is lacking, but platelet transfusions are largely given in trauma, cardiac surgery, treatment of leukemia and other hematologic cancers, bone marrow transplantation and some solid organ allografts, such as liver transplants.

4

Table 4.1. Blood transfusion statistics

Annually, four million recipients in the United States

- 12 million units red blood cells

- 4.7 million units of platelets

- 600,000 Apheresis platelets

- 2.3 million units of plasma

- 1 million units cryoprecipitate

Approximate red blood cell transfusion rates: (per year)

Northeast U.S.	50-60 per 1000
Midwest	35-50 per 1000
Northwest U.S.	35-45 per 1000

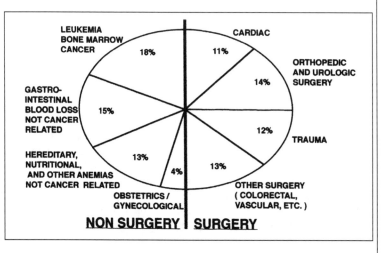

Fig. 4.1. Red cell transfusion recipient subpopulation (USA).

Informed Consent and Explanation of Blood Options

One of the most important and difficult areas in clinical transfusion medicine is the question of informed choice [consent] for blood transfusion and an adequate explanation to potential blood recipients of blood options (All Blood Options).

In understanding informed choice, it is important to appreciate the following: (1) There is a need to discuss the risks, benefits, and alternatives to blood transfusion and to ensure and that the potential recipient or his/her representative has an opportunity to ask questions. The rationale for this arises for two reasons: (a) there are *real risks* associated with blood transfusion. These are the risks of "immediate" death, which can occur with acute hemolytic reactions, acute bacterial infections or rarely anaphylactic reactions. Such real risks are now extremely uncommon, with a frequency of occurrence of less than 1:50,000, and death resulting, in probably less than 1:500,000. There are other risks associated with transfusion, such as minor allergic and febrile reactions, but these result mainly in patient discomfort and short-term morbidity (Chapter 32). These events are more frequent, and, with platelets, may occur in 3-20% of transfusions. The overall frequency is approximately 1%. The likely transmission of any viral disease causing significant morbidity currently is certainly less than 1:3,000 and likely less than 1:34,000. Therefore, in evaluating the real risks associated with allogeneic blood transfusion for red blood cells there are actually no risks of any high frequency (> 1% per unit). On this basis, it has been argued that the need to discuss specific risks of blood transfusion is questionable. (b) The more important reason to discuss risk in blood transfusion is on account of *perceived risks*. Inherent in this concept is that the patients' perceived risks might be instrumental in making an informed choice. For example, although the per unit risk of transfusion-associated HIV-1 is less than 1:225,000 (and perhaps as low as 1:1,000,000), an individual's fear or concern regarding HIV-1 infection may be such that they may not wish to receive any allogeneic blood. Statistically, this can be discounted, but nevertheless, it is of importance to the person as an individual. Each individual's risk tolerance for undesired low frequency events may vary and cannot be assumed. In summary, no real risks of 'high' frequency are known to exist with transfusion, but discussion of febrile-type reactions or urticaria would seem reasonable. In addition, discussions regarding real risks of *very* low frequency (< 1:1,000) are important in order to lessen patients concerns. Discussion of benefits should be simple such as improved sense of well being and more energy with red blood cells or prevention of bleeding with platelets or plasma.

Clinical Transfusion Medicine, by Joseph D. Sweeney and Yvonne Rizk. © 1999 Landes Bioscience

Having established a rationale for discussing the risks and benefits of blood transfusion, it is then important to have a consistent *process*. There are a number of ways by which this can be achieved. Providing patients with printed or audio-visual material in a timely manner in relation to the transfusion may be helpful (these materials can be provided to custodians if the patient is unable to make such an informed choice). It is important, however, that any materials made available should be in a format that is comprehensible to the patient; for example, the language used should be simple in order to facilitate comprehension and a language other than English may be appropriate in certain geographical locations. These materials, however, are only a supplement to, and not a substitute for, a discussion with the patient or with the patients' representative, by a physician or some other informed discussant, such as a nurse, nurse practitioner, or a physicians' assistant. The need for a blood transfusion should first be outlined. Possible risks should then be discussed with alternatives if relevant. An opportunity needs to be offered to the patient to raise any questions or concerns, they may have regarding transfusion.

There are various mechanisms by which the informed choice process can be documented. (a) In the simplest sense, the physician prescription or signature for the blood product could constitute, in itself, the documentation. This is the mechanism by which documentation is occurring by default in these institutions in which there is no specific reference to blood transfusion in any of the formal consent documents. (b) A written notation can be made in the patients progress notes, stating that the patient consented to blood transfusion. The potential problem in this situation is for different types of notes to be written by different physicians raising concern regarding the consistency of the information presented to the patient. Simplifying the progress notes to make a standard statement such as "risks, benefits, alternatives or blood transfusion were discussed with the patient and an opportunity was offered to present questions" might be the best approach, followed by the signature of the physician and/or other health care personnel. (c) Institutional consent forms where specific reference can be made to transfusion. In such cases, a patient's signature could attest to the consent to blood transfusion. There may be further detail regarding risks, benefits, alternatives or, uncommonly, some actual data quantifying these risks. (d) A different kind of consent form is a procedural consent form, (for example, before surgery or a medical procedure such as a liver biopsy). Procedural consent forms, particularly surgical consent forms, will often make specific reference to the possible need for blood transfusion. This is appropriate, as more than 50% of all blood transfusions, occur in association with surgery. The degree of documentation on this consent form may vary from simple single sentences to an extended paragraph. (e) The most contentious manner of documentation is the use of a specific consent form for blood transfusion. This usually outlines the risks, benefits and alternatives in a clear format separating more common from very uncommon risks. This consent form offers the best form of documentation. The difficulty with specific consent forms, is that a considerable amount of time and effort is spent in getting additional signatures in a situation where the patient or the patients representative is already overburdened emotionally and physically. In addition, failure to obtain this consent (noncompliance) could paradoxically increase legal exposure. Nevertheless, many institutions have elected to establish a transfusion specific consent form. It is also been suggested that the existence of a transfusion-specific consent form is more likely to result in a greater compliance by physicians in discussing risks and benefits with patients, although there is no actual data that this is the case. Both the *process* of informed choice and the *documentation* of informed choice should be audited periodically to ensure compliance. These concepts are shown in Table 5.1.

Related to the concept of informed choice is the question of all blood options (ABO). These constitute possible alternatives to allogeneic blood products and are itemized in Table 5.2. Blood options, or alternatives, are an important part of the consent process. Inherent is the option to refuse blood transfusions, most commonly, for example, with Jehovah's witnesses and Christian Scientists. It is important to document such refusal on a consent form, if possible, or in a progress note, in order to prevent inadvertent transfusions and reduce liability. Pharmacological alternatives or augmenters are available, which in some situations are appropriate (Chapter 23). The option for autologous blood (see Chapter 3) requires careful discussion with appropriate patient populations. The potential use of directed donors should neither be encouraged or discouraged. This is because directed

---

Table 5.1. Considerations regarding informed choice for blood transfusion

1. Rationale for discussing risks, benefits and alternatives
   a) Real risks
   b) Perceived risks

2. Process
   a) Discussion with patient by physician, or informed discussant (nurse, nurse practitioner, physician assistant, etc.)
   b) Printed or audio-visual materials made available for recipients

3. Documentation
   a) Physician signature prescribing the blood product
   b) Patient progress notes
   c) Patient signature on:
      i)   General (or institutional) hospital consent
      ii) Procedural consent
      iii) Transfusion specific consent

4. Consistency of Process and Documentation

5. Audit of Both Process and Documentation
   a) Patient survey (process)
   b) Chart survey (documentation)

donations (Chapter 2) are not known to be safer than the standard volunteer donations although the patient may perceive them as such. The physician therefore, should avoid expressing any bias or opposition to the use of directed donors. Lastly, as patient populations become more educated, there may be discussions concerning the use of specialized blood products. Examples of these specialized blood products are the use of leukocytoreduced blood products in colorectal surgery (Chapter 13), which has shown in some studies to reduce postoperative infectious rates and reduce length of stay or virally attenuated plasma, which is prepared from large pools and subjected to chemical treatment (Solvent-Detergent or SD plasma). This product may be inherently safer in that it is less likely to be associated with viral disease transmission. However, the use of a large pool size is worrying, since the SD process does not inactivate nonlipid-enveloped viruses (Chapter 29). A newer form of fresh frozen plasma, called fresh frozen plasma, donor retested (FFP-DR) is also likely to be available. This is plasma which has been quarantined until the donor returns to donate. These products may allay the concerns of some transfusion recipients.

5

Table 5.2. All blood options

---

1. Option to Refuse Blood Transfusion
    a) Jehovah's Witness
    b) Christian Scientists

2. Option to Seek Pharmacologic Alternatives or Augmenters
    a) DDAVP (Hemophilia A; von Willebrand's disease)
    b) rh-Erythropoietin (renal failure; pre-elective surgery)
    c) Estrogens (renal failure)
    d) Recombinant FVIII:C or IX:C

3. Option for Autologous Blood
    a) Predeposit Autologous Blood
    b) Preoperative Hemodilution or Apheresis
    c) Intraoperative Salvage
    d) Postoperative Salvage

4. Option for Directed Donors

5. Option for Specialized Products of Potential Lower Rrisk
    a) Leukoreduced Blood Products
    b) Virally Attenuated Plasma
    c) Fresh Frozen Plasma, Donor Retested

The ABO and Rhesus Systems

Although there are now in excess 26 known different blood group systems identified with an associated 254 separate antigens on human red cells, only the red cell antigens within the ABO system and a single antigen (D) within the Rhesus system, are routinely assessed. Antigens in the other blood group systems are only assessed in certain circumstances. These blood group systems, known as minor blood group systems, will not be discussed further except in some specific situations for clarification (Chapter 17; 32) and interested readers should refer to more comprehensive textbooks.

The discovery of the ABO system at the end of the nineteenth century laid the foundation for clinical transfusion practice. It is now known that ABO antigens expressed on red cells are determined by genes located on the long arm of chromosome nine. These genes code for glycosyl transferases, which attach different carbohydrates (sugars), to a terminal galactose of an oligosaccharide chain. These oligosaccharide chains are attached to phospholipids in the red cell membrane and to proteins (glycoproteins) in plasma. The ABO system is illustrated in Figure 6.1. In blood group A, the terminal sugar is N-acetyl D-galactosamine and in blood group B, D-galactose. Individuals of blood group AB contain both A and B antigens on their red cells. Individuals of blood group O lack a functional transferase and hence do not transfer either sugar of type A or B. The distribution of the different ABO types differs substantially between different populations as illustrated in Figure 6.1. Prominent is the relatively larger proportion of Group B in African-Americans and Asians.

Within the ABO system, individuals who lack the A or B antigens have the reciprocal antibody present in plasma. This antibody develops early in life, probably due to exposure to bacterial cell walls, which contain oligosaccharides and provide the stimulus for specific antibody formation. The ABO system is expressed on all cells and soluble ABO antigens are present in plasma (regardless of secretor status). In rare circumstances a discordance exists between the antigens present on the surface of red cells and the reciprocal antibodies in plasma. This is called an ABO discrepancy and occurs in unusual situations, such as subgroups of A (A_3), immunoglobulin deficiencies, the presence of cold agglutinins, rare blood types, or unexpected antibodies outside of the ABO system. These can usually be resolved easily by the blood bank but can cause a delay in the availability of red cells for transfusion. The overwhelming importance of the ABO system lies in the potential for acute intravascular hemolytic reactions to occur with fatal outcome should an incompatible reaction occur (Chapter 32). It is for this reason that the most critical aspect of the red cell compatibility testing is determination of the ABO system of the recipient and ensuring compatibility with donor blood.

Clinical Transfusion Medicine, by Joseph D. Sweeney and Yvonne Rizk. © 1999 Landes Bioscience

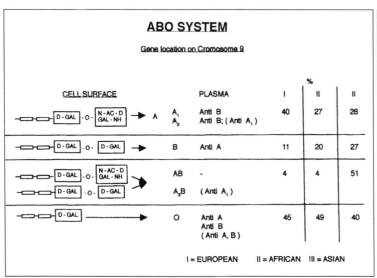

Fig. 6.1 ABO blood system

Next, in importance to the ABO system is the Rhesus system. The Rhesus system differs biochemically from the ABO system. Expression of the Rhesus system is determined by two genes located closely together on chromosome one. One of these genes codes for a protein which expresses the important antigen called D. Individuals expressing this antigen are called Rhesus positive, (or Rh (D) positive) and those who lack the antigen, called Rhesus negative (or Rh (D) negative). The second gene codes for over 50 antigens within the Rhesus system (CE gene). Thus, a Rhesus negative person, while lacking D, does express Rhesus antigens and the terminology is, therefore, unfortunate. However, the terminology is unlikely to change. The concept of the Rhesus blood group system is illustrated in Figure 6.2. The Rhesus system antigens are expressed on a lipoprotein molecule present in the red blood cell membrane. Antibodies to Rhesus usually occur only with immune stimulation such as pregnancy or a previous transfusion, and spontaneous occurrence is rare. The antigen D within Rhesus is further complicated in that some individuals have fewer D antigen sites on the red cell membrane (called weak D) and others have an abnormal type of D (partial D). For transfusion purposes, weak D recipients will often type as Rhesus negative unless a more sensitive test is performed and will receive Rhesus (D) negative blood (Weak D donors will be typed as Rhesus positive by a blood center). However, partial D individuals usually type as Rhesus (D) positive. In some cases, they develop an antibody to the abnormal part of the D antigen, giving rise to an apparent paradoxical situation where a Rhesus positive subject develops anti-D.

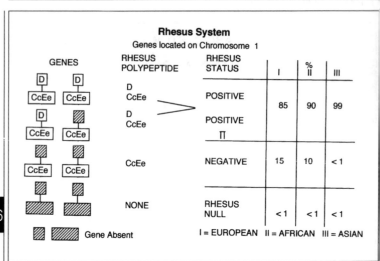

Fig. 6.2. Rhesus blood system

Antibodies within the ABO system tend to be IgM and are often complement fixing, causing massive intravascular hemolysis. Antibodies to the Rhesus system antigens, (e.g., anti-D) are generally IgG and rarely, if ever, fix complement. They cause predominantly extravascular hemolysis (Chapter 32), and thus, hemolytic reactions to Rhesus system antigens tend to be clinically mild relative to ABO system antigens. The reason for D typing within the Rhesus system is because of the potential to form antibodies to D. A unit of D positive blood transfused to a D negative individual, will result in the formation of anti-D in 70-95% of recipients and this is best avoided. This is of particular importance in females of childbearing years, as this antibody can cause severe hemolytic disease of the newborn (Chapter 24). It is because of this high propensity to form anti-D by immune challenge, that blood is routinely tested for the D antigen. As in the case of the ABO system, there are significant differences in the expression of different antigens in different populations. The most obvious difference is the very low prevalence of Rhesus negative (D negative) subjects among Asian populations. Also, Rhesus antigens differ from ABO in that expression is restricted to the red blood cell membrane.

Compatibility Testing and the Importance of Proper Recipient Identification

Compatibility testing is the major process which separates the transfusion of red blood cells from the administration of other types of pharmaceutical products (Chapter 1). The purpose of compatibility testing is the avoidance of a *hemolytic reaction*. This is of cardinal importance, since some types of hemolytic reactions can result in a fatal outcome. The technical procedures used in compatibility testing have undergone significant changes over the past few decades. The critical steps in this process are shown in Table 7.1.

Although not often properly appreciated or given adequate emphasis, the first important step is correct identification of the intended recipient at the bedside at the time of venipuncture. This requires accurate labeling of the specimen tube by the phlebotomist *before leaving the site* at which the venipuncture was performed. This is interpreted, at a minimal, as the first name, last name and a patient specific identifier, such as a medical record number, social security number, date of birth, etc. Unique identifiers are preferred, but may not be practical. The phlebotomist should be identifiable, either by a complete signature or a first initial and a last name; initials only may be acceptable. In addition, the time and date of sample collection should be written on the tube. All of this data is important and the only information which could be subsequently added, if omitted, would be the time or date of draw, since this information does not impact on identification. The specimens may be collected into prelabeled containers (tubes). If this is the practice, it is particularly essential for the phlebotomist to sign the tube immediately after sample collection. It is preferable, but not required, to collect the blood sample into an unlabeled container. The sample must then be immediately labeled at the bedside by affixing a preprinted label or handwriting the information. Labeling, and the identification of the phlebotomist, must be completed at the site of collection (bedside). It is unacceptable to complete any part of this process at a different location from the actual site of collection. Errors at the point of sample collection set the stage for the fatal outcome of a subsequent transfusion.

The accuracy of this information is important for an additional reason. All Blood banks check for previous records on a patient and accuracy of the information allows early identification of a potential problem or delay in blood availability.

The next most important step is the ABO typing of the specimen. This is performed by identifying the ABO antigens on the red cell of the intended recipient (called a front type or forward type) and by identification of antibodies to ABO antigens in the serum or plasma (called back type or reverse type). In general, as

---

Table 7.1. Important steps in compatibility testing

---

1. Correctly identify the intended recipient at the time of *venipuncture*. Label the tube at the *site* and *time* of collection. Sign the tube verifying confirmation of identification.

2. ABO typing of specimen.

3. D typing (Rhesus) of specimen.

4. Screening the serum for unexpected antibodies (called screening or; indirect coombs or; indirect antiglobulin test).

5. (a) If #4 is negative (normal), linking the ABO type of the donor unit with the ABO type of the intended recipient.

   (b) If #4 is positive (abnormal), linking unexpected antibodies in the recipient with antigen negative donor units.

6. Correctly identify the recipient at the time of blood administration.

7

indicated in Chapter 6, front typing and reverse typing will give complementary data. Although the need for both front type and reverse typing could be questioned, consistency in the front and reverse type give a sense of security with regard to the correct identification of ABO type. Rarely inconsistencies, as stated in Chapter 6, can occur but these can usually be resolved by the blood bank. In urgent situations, transfusion of Group O red cells and AB plasma is preferred until the discrepancy is resolved.

The next important step is typing for a single antigen, called D, in the Rhesus system. Individuals who type as D positive are called *Rhesus positive* and those who type as D negative, are called *Rhesus negative*. This is a front type (antigen type). No "back typing" is performed for the Rhesus system, since unlike the ABO system reciprocal antibodies are not routinely detected in the serum. Antibodies to D or other Rhesus antigens are detected in the antibody screen (see below).

The next step in compatibility testing is screening of the serum or plasma for unexpected antibodies outside of the ABO system. This is performed using either two or three sets of group O red cells. This test is commonly called an antibody screen. Other names that describe this test are the "indirect coombs test" but in the laboratory, this is called the indirect antiglobulin test (IAT). In this test, the serum or plasma from a potential recipient is incubated with the screening cells. Various temperatures of incubation and additional reagents may be added before examination for agglutination in a final phase in which an antiglobulin reagent, e.g., (rabbit antihuman IgG) is added to detect sensitized red blood cells (red cells coated with an antibody present in the recipient's plasma/serum). Between 1-8% of potential blood transfusion recipients may show the presence of an unexpected antibody. These antibodies may be directed against antigens in the Rhesus system (often anti-D in Rhesus (D) negative patients), but also against antigens of the so-called minor blood group systems. For a more detailed description on these antibodies, the reader is referred to reference textbooks on blood transfusion.

If the antibody screen is negative (the common situation), then the next step is to link the ABO type of the donor unit with the ABO type of the intended recipient. A simple cross-match (called immediate spin) can be performed at this point which ensures ABO compatibility or computer records of the ABO types can be used to match the donor and recipient (electronic cross-match). If the antibody screen is positive, identification of the antibody must be performed, and transfusion of red blood cell units which lack the antigen is required. In these cases, a more extensive cross-match (antiglobulin cross-match) procedure is performed, which is similar in principle to the antibody screen. This accounts for the delay often encountered in the availability of blood for these patients.

After these steps have been completed, the blood may be dispensed for transfusion. A most important part of compatibility is correct identification of the recipient at the time of blood administration. There may be different procedures by which this is achieved, depending on the site of transfusion. In the past, most blood was transfused in a hospital setting. Hospital patients tend to be more easily identifiable, since they commonly have an identification band attached which contains identifying information. This is not always the case, however for example, in emergency rooms, outpatients or in other settings, such as operating rooms when the identification bands may be removed (e.g., for A-line insertions) or inaccessible. Different protocols for proper recipient identification should be in place for each location to ensure that the recipient is properly identified. Although identification of a recipient by one individual is acceptable, this is most commonly performed where possible by two individuals, one of whom is generally either a nurse or physician. This may not be possible, however, in all locations. Errors at this point are fortunately uncommon but can result in very severe reactions, occasionally with fatal outcome. This is a particular problem in situations where blood is being transfused under stressful conditions, such as rapidly bleeding patients in the operating room or trauma patients in emergency rooms. This problem is further compounded in that such patients may be unconscious, and hence the early clinical features of hemolytic reaction may go undetected. Blood observed in a urinary catheter or severe hypotension may be the first indication of acute hemolysis. Each of these sites, therefore, may need to develop specific procedures to ensure that proper identification occurs.

Figure 7.1 illustrates the relative importance of the technical tests in compatibility testing. Transfusion of blood from a Caucasian donor population to a Caucasian recipient without regard for ABO type of the recipient or donor blood would result in an ABO incompatibility in about 33% of cases. If the ABO system is matched appropriately, and no other tests are performed, then the likelihood of a successful transfusion reaction is high. The common problem would be the transfusion of Rhesus positive blood to Rhesus negative individuals, which would result in the formation of anti-D in approximately 90% of Rhesus negative recipients.

Transfusion

Without Regard to ABO or Rhesus (D)	33% Catastrophic Outcome
ABO Compatible Without Regard to (D)	12% Mild or Undesired Outcome
ABO Compatible Rhesus Compatible	3% Mild Reactions Occasionally Severe
ABO Compatible Rhesus Compatible Antibody Screened	< 1% Mild Reactions

Fig. 7.1. Different outcomes from red cell transfusions when different technical steps in compatibility testing are performed.

This would cause difficulties with subsequent transfusions or future pregnancies. If ABO typing and Rhesus typing are performed, then the likelihood of an uneventful transfusion episode is high. The difficulty in these cases would be the small number of transfusion recipients, (overall less than 3%) who would have an unexpected antibody, which reacts with an antigen in the donor blood. With antibody screening, nearly all of these antigen are eliminated, with the result that in less than 1% of all transfusions are normally associated with reactions. These are generally of a mild nature causing mostly short-term patient discomfort. These reactions are unrelated to hemolytic reactions and mostly due to contaminating white cells, and are discussed in more detail in Chapter 32.

It is sometimes thought that compatibility testing is synonymous with the term "crossmatching". Compatibility testing has both clerical and technical procedures, and only one of these procedures is the crossmatch. In most Blood banks, the crossmatch remains a technical procedure (actual testing of recipient serum/plasma with donor red blood cells) but increasingly the crossmatch is becoming a clerical function (electronic crossmatch) for those patients with negative antibody screens. The difficulty in conceptualizing compatibility testing as synonymous with crossmatching is that it undermines the enormous importance of the clerical identification steps at the time of *sample collection* and at *the time of blood administration*. In addition, if only the ABO type and Rhesus type are matched, the antibody screening is negative and no technical crossmatch performed, the likelihood of

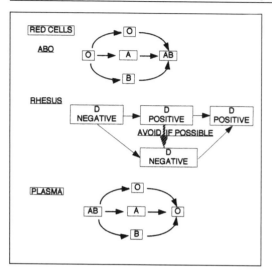

Fig. 7.2. Safe compatibility choices when ABO and Rhesus identical blood components are not available.

any significant hemolytic reaction is statistically, extremely low. Thus, crossmatching as a laboratory technical procedure is being de-emphasized, and in the near future, clerical (electronic) crossmatching is likely to become standard for the majority of blood units transfused.

Lastly, it is important to appreciate the distinction between ABO and/or Rhesus identical blood and ABO and Rhesus *compatible blood*. The transfusion of various blood products to transfusion recipients which are compatible is demonstrated in Figure 7.2. In general, it is common practice to transfuse ABO and Rhesus identical blood where possible. However, under certain circumstances, it is not uncommon to transfuse ABO compatible blood, such as group O red cells to non-group O recipients or group A red cells to blood group AB. In addition, Rhesus negative units may safely be transfused to Rhesus positive patients, but this is uncommon because of the limited availability.

The transfusion of a Rhesus positive unit to a Rhesus negative recipient is best avoided. However, shortages of Rhesus negative blood do occur, and this occasionally has to be done to conserve community supplies. The recipients in these cases should be preferably males or females well beyond childbearing age (arbitrarily > 50 years) and have no evidence of an anti-D (negative antibody screen). The inadvertent transfusion of a Rhesus positive unit to a Rhesus negative female of childbearing age is a major error and constitutes the most important practical aspect of ensuring Rhesus compatibility for most recipients.

The Administration of Blood Products

The administration of blood products requires proper compliance with a written procedure, the important elements of which are outlined in Table 8.1.

First is proper recipient identification and ensuring the compatibility of the product. For red cell transfusions, both ABO and Rhesus compatibility should be ascertained. If there are any questions at this point they should be immediately addressed to the blood bank for clarification. Under certain circumstances, non-identical ABO blood will be administered to patients, for example, blood group O red cells to non-O recipients or blood group A red cells to AB recipients. In addition, Rhesus negative products may be safely transfused to Rhesus positive patients, and on occasion, when Rhesus negative shortages exists, Rhesus positive units may knowingly be transfused to certain groups of Rhesus negative patients. When blood is dispensed from a blood bank, a record is attached to the bag. This record contains information identifying the blood in the container (ABO, Rh and unit #) and information identifying the intended recipient (name, medical record #, other identifiers). This record, therefore, links the suitability of the blood in the container with the recipient. Confirming the correctness of this information at the bedside may be the last opportunity to avert a severe hemolytic reaction.

Inspection of the blood bag for leaks and the general appearance of the product is important to detect contamination of the product with bacteria or other substances. The administration set should have an in-line filter; and routine intravenous infusion sets for fluids are not acceptable. This filter removes particles with an average size of between 170-260 microns (μ). Blood administrations sets commonly have both a drip chamber and a filter chamber, the former allowing the calculation of the rate of administration of blood and the filter chamber ensuring the removal of debris which may have accumulated during storage. The drip chamber allows 10 and 20 drops per minute (10 drops = 1 ml) and the transfusionist can calculate the rate of transfusion and likely duration.

Under some circumstances, the rate of blood transfusion can be increased by the use of either a pressure cuff or an electromechanical device, such as a pump. Although large pressures may be applied with a pressure device, this is not known to be harmful to either red blood cells or platelets. The major concern with pressure cuff devices is either bag rupture or the potential for air embolism. When pumps are used routinely for red cell transfusion, the manufacturer should have information on file that hemolysis of red cells does not occur during normal operation of the device. Pumps can also be used to transfuse platelets, particularly in a pediatric setting. In general, these pumps have not been shown to alter platelet function. Thus, use of electromechanical devices is acceptable practice for the transfusion of blood products. Pumps also allow a greater degree of control of the rate of transfusion than might be possible by visual counting of the number of drops.

Clinical Transfusion Medicine, by Joseph D. Sweeney and Yvonne Rizk. © 1999 Landes Bioscience

---

Table 8.1. Important steps in blood administration

---

1. Ensure proper recipient identification, *ABO* compatibility *and Rhesus* suitability *of the product.*

2. Inspection of the blood bag for product appearance and any leaks.

3. Ensure that the administration set has an in-line filter.

4. Do not add to or infuse blood with any fluid or medication, other than 0.9% saline.

5. If a mechanical pump is used routinely, information regarding lack of hemolysis is appropriate.

6. If blood warmers are used, these should be quality controlled at least semi-annually, or more often, depending on use.

7. Vital signs should be taken before the transfusion.

8. The initial rate of transfusion should be slow (about 1-2 ml/minute) to detect and respond to sudden severe unexpected events, i.e., acute hemolysis, bacterial sepsis, or anaphylaxis.

9. The duration of a red cell transfusion is optimally $1^1/_2$ hours, but should not exceed 4 hours.

10. Vital signs should be taken after the transfusion or at any time if a reaction occurs.

11. If a reaction occurs, stop the transfusion, maintain an open IV line with saline and evaluate (Chapter 32).

12. Avoid sampling from or above the IV site during, or immediately after, the transfusion.

13. If the transfusion is uneventful, discard the empty bag in a manner consistent with the disposal of biologic waste.

Blood is sometimes transfused using blood warmers. It is rarely necessary to transfuse red cells using a blood warmer when the duration of the transfusion is in excess of 1 hour, the only possible exception being recipients with cold agglutinins. Platelets are stored at room temperature, and other products such as plasma and cryoprecipitate are thawed at 37°C. However, blood warmers are used in the operating room, or in patients with cold agglutinins, or in massive trauma when blood needs to be transfused rapidly, (50-100 ml/min). Particular attention needs to be paid to the quality control of these blood warmers, at least on a quarterly basis, if in frequent use, particularly that excessive temperatures do not occur. When red cells (preferably less than 42°C) are exposed to temperatures higher than 42°C, hemolysis may occur.

With red cells, the initial rate of transfusion should be set at 1-2 ml/min, for approximately 15 minutes. This is to detect and respond to any sudden or unexpected clinical events such as acute hemolytic reactions, bacterial sepsis or anaphylaxis. Although it is not uncommon practice to measure vital signs at this time, simple questioning or observation of the patient as to whether they are experiencing any discomfort is adequate. After this time, the rate of transfusion can be increased in order to complete the transfusion over a period of 1-2 hours. In some institutions, it is practice to routinely transfuse a unit of blood over a period of 4 hours. This is, of course, acceptable, but it is not required, and may be inconvenient. For other blood products, such as plasma or cryoprecipitate, the rate of infusion should be set to meet the desired clinical objective and be consistent with the patient's tolerance for increased intravascular volume. Platelet transfusions are often administered more rapidly, over a period of 15-30 minutes. Such rapid platelet transfusions can occasionally result in the occurrence of febrile or urticarial reactions in the patient. The occurrence of fever in association with platelet transfusion should keep the transfusion alert to the possibility of bacterial contamination. Therefore, close observation is always appropriate for platelet transfusions whenever such rapid infusions are performed. If a reaction occurs, the critical event is to stop the transfusion, maintain the intravenous line open with saline and evaluate the clinical situation (see Chapter 32). Vital signs should always be taken immediately if a reaction occurs and are required to be taken routinely in the U.S. after completion of an uneventful transfusion. If the transfusion is uneventful, the empty bag may be discarded immediately. However, some institutions retain the bag for a period of 6-8 hours, since rarely a reaction can occur up to several hours after completion of the transfusion.

No fluid or medication other than 0.9% saline should be added or connected in any way to the administration sets in which human blood products are being transfused. The use of solutions in surgery such as Ringers lactate, which contains calcium, may cause small clots to form and other fluids and 5% dextrose can result in hemolysis. In addition, sampling should be avoided from the IV site used for transfusion in the period during and immediately after a transfusion. Red cell products have an Hct of 55-60 and could cause an erroneous blood count result Stored blood contains high concentrations of potassium (30-50 mEq/L) and glucose (300-500 mg/dl) which may cause confusion in the interpretation of chemistry tests.

Blood Transfusion in Surgery I: Ordering Practices and Transfusion Styles

Approximately 50% of all red blood cells are transfused in association with surgical procedures, many of which are elective in nature. On account of this large percentage, the transfusion practices of anesthesiologists and surgeons greatly impact on the blood resources of the community.

Ordering practices are those practices which relate to the anticipated or potential use of blood in association with surgery or invasive diagnostic procedures. Mostly, these develop on the basis of historical clinical experience with the procedure being performed. As shown in Table 9.1, there are various potential approaches to ensuring the availability of blood in the event of hemorrhage. This reflects nothing more than a hierarchy of probabilities that any allogeneic blood may need to be transfused. First, those situations where the blood use is exceedingly rare are unlikely to benefit from any blood-banking test for compatibility. Examples of these kinds of procedures are superficial skin biopsies or lumpectomies. In the past, specimens were routinely sent to the blood bank for typing, or screening, but this is wasteful. The next level is blood typing only, but this is of little value, as the patient's blood type has no diagnostic value in surgery. If blood is needed in an emergency, ABO identical blood could be issued, but this is no known gain in safety over the emergency issue of group O blood. A third level of request is the so-called "type and hold". This does not generally increase safety, since if blood is needed urgently, it will simply be issued as ABO identical or group O, i.e., similar to a "type only" request. A fourth level of request is "type and screen". This is a *very useful request* in situations where blood may (occasionally) be needed. From a practical point of view, this approach should be used for the majority of such surgical procedures. When a type and screen is requested, the ABO and Rhesus (D) type is determined and the serum screened for unexpected antibodies (see Chapter 8). A variation of type and screen is to screen for unexpected antibodies but not to type the patient ("screen and hold"). This is an interesting approach in the management of situations where blood transfusion is rarely required. If the antibody screen is negative, the transfusion of group O uncrossmatched blood has almost no statistical likelihood of a hemolytic reaction. Screen and "hold" is an uncommon request as most blood banks discourage performing a screen without a type and therefore "type and screen" is the more common approach.

For those procedures however, in which blood is commonly transfused, the approach is to type, screen and crossmatch (or have available electronically) a predetermined number of units sometimes called "type and crossmatch". Under

---

Table 9.1. *Ordering practices: anticipated or potential use of blood*

1. *No specimen*: Suitable when blood use is exceedingly rare.

2. *Type only* (ABO, Rhesus): A practice of no known value.

3. *Type and "hold"*: Better to request #4 or consider #1, depending on the procedure.

4. *Type and Screen*: Suitable when blood use is occasional.

5. *Screen and Hold*: This is a reasonable approach if blood use is *very occasional*. However, blood banks have a bias to type always and probably #4 is preferable.

6. *Type, screen and crossmatch*: Suitable when blood use is common or routine.

---

these circumstances, compatible blood is identified and set aside for potential use, usually for a 48 or 72 hour period. There is no clear definition of what is considered "commonly transfused" but, in general, if blood is transfused in more than 50% of cases for any given surgical procedure, it is not unreasonable to have crossmatched blood available. The concept of crossmatching has undergone significant evolution, however. Patients with negative antibody screening (97% of specimens, Chapter 8) can now receive ABO identical blood dispensed without a technical procedure being performed (electronic crossmatch). This greatly expedites the availability of red cells in the event of unexpected hemorrhage. In the past, there has been a trend to over request crossmatched blood in order to give a "cushion" in the event of unexpected hemorrhage. This approach results in unnecessary crossmatches and a high crossmatch to transfusion ratio (CT Ratio). In situations where the antibody screen is positive, the blood bank commonly doubles the number of units made available (crossmatched) as a matter of practice. Therefore, the practice of over ordering crossmatched blood because of concern surrounding the potential inability of the blood bank to respond to unexpected situations should not be justifiable. Most over-crossmatching of blood has evolved as a perception issue on the part of operating room personnel that the blood bank will be unable to respond to an emergency situation. Therefore, development of good communication between the transfusion service, anesthesiologists and surgeons is critical in overcoming this perception.

On account of this, most institutions develop what is described as a maximum (surgical) blood ordering system or MBOS. This is a schedule where the number of units to be crossmatched, if any, are agreed by the surgical staff and a written list is assembled. When the MBOS is implemented, there tends to be a significant reduction in the amount of blood that is routinely crossmatched. The MBOS list should ideally show three types of procedures: (a) These procedures for which a specimen is not required, (blood almost never transfused), (b) type and screen, only (blood rarely transfused) and (c) type and crossmatch for a predetermined number of units (blood commonly transfused). The surgical procedures can be arranged by surgical service, alphabetically, or procedural code. At the time of sample collection (if appropriate), the request should indicate the type of surgical procedure and surgical code (e.g., CPT code or other). This can then be translated into a type and screen, or type and crossmatch, by the blood bank staff.

Related to ordering practices for blood transfusion is decision making regarding transfusion. This is often called "transfusion practices" or "transfusion styles". Transfusion practices and styles tend to evolve on the basis of empiric clinical experience and not on the basis of clinical studies. Transfusion styles differ from transfusion practices, but have in common their origin in empiric clinical experiences. Transfusion styles often have developed from unanalyzed, partially analyzed, and occasionally anecdotal experiences. Table 9.2 shows important distinctions between transfusion practices and transfusion styles. Both can result in either over use or inappropriate use of blood transfusion, but, also potentially, *under use* of blood transfusion. The most important difference between transfusion practices and transfusion styles is the ability to effect intra-institutional change. Transfusion practices evolve on the experience of a physician or group of physicians within an institution. They are left unchanged until challenged with data or educational material. Under such circumstances, these practices can be changed, resulting in a better utilization of blood products. Transfusion styles differ, however. Transfusion styles, although possibly based initially on empiric, often anecdotal, clinical experience, are often reinforced by the culture of a department within

Table 9.2. Importance of differentiating transfusion practices from transfusion styles

Transfusion Practices	Transfusion Styles
1. Develop/evolve within the framework of empiric clinical experience	Develop/evolve within the framework of empiric clinical experience or tradition, sometimes anecdotal
2. Determined by individual physician or group experience	Institutionally determined by culture or attitude
3. Often amenable to change by logic, hard data and education	Resistant to change. Short term changes revert to old styles. Logic/data viewed skeptically. Change requires behavioral adjustment
4. New physicians on staff may influence practices and cause change	New physicians on staff 'adapt' to the transfusion style (sometimes reluctantly)
5. May result in product wastage	Often results in product wastage

an institution. They tend to be resistant to change. Educational intervention sometimes causes short-term changes, but reversion to the old transfusion styles tends to recur. New physicians on staff are frequently capable of changing transfusion practices. However, new physicians on staff tend not to influence transfusion styles; and adapt, in time, to the style of the institution. Questionable transfusion practices and transfusion styles result in considerable blood product wastage and unnecessary cost, reducing the available blood supply within the community.

Illustrative examples of transfusion styles are (1) the routine administration of plasma in association with red cell transfusions in surgery. In the past, surgeons or anesthesiologists would transfuse a unit of plasma for every two or three units of red cells transfused during surgery. For most patients with normal hemostatic mechanisms presurgically, there is no evidence that this is of any benefit. Transfusion of plasma may, however, be useful when large volumes of allogeneic red cells or salvaged autologous red cells are transfused (approximating, 0.5-1 blood volume) and initial replacement is red cells in crystalloid. (2) The routine transfusion of platelets presurgically, if the platelet count is less than 100×10^9/L outside of the context of neurosurgical or ophthalmic procedures. In clinical situations where the operative field is well visualized and hemostasis can be controlled by good surgical technique, this practice is of no known benefit. Patients who exhibit excessive microvascular oozing with platelet counts less than 50×10^9/L, may, on the other hand, benefit from platelet transfusions. (3) The routine transfusion of red blood cells to patients with a hemoglobin below 10 g/dL. There is no empiric justification for this approach which, until recently, was largely unchallenged. Some patients, however, may indeed, benefit from transfusion if the hemoglobin is less than 10g/dL in situations where the clinical circumstances indicate critical organ ischemia, and there is risk of imminent hemorrhage (Chapter 26).

The importance of ordering practices, transfusion practices and styles cannot be overemphasized. The ability of the transfusion service to function adequately to meet the surgical needs and promote the optimal usage of blood resources in a community are significantly jeopardized by inappropriate institutional practices or transfusion styles. Much of clinical transfusion medicine is concerned with understanding these practices and styles and intervening to effect a change to better transfusion practice.

Blood Transfusion in Surgery II: Cardiac and Vascular Surgery

Blood transfusion in cardiac surgery accounts for 10-14% of all red cells transfused in the United States. Mean usage/patient is about 5 units, although there is a huge variation between different institutions. This results in 1.2 million units per year transfused in the United States. Transfusion practices in cardiac surgery are, therefore, of great importance to hospital blood banks.

The cause(s) for this variation in practice is not entirely clear, but current evidence indicates that certain kinds of patients have an increased likelihood of blood transfusion. Female gender, increased age (over 70), low preoperative hematocrit and extensive procedures such as combined bypass and valve procedures with long pump runs are predictive of increased blood usage. Other determinants of blood use appear to be choice of the vascularization vessel, either saphenous graft or internal mammary graft. Even allowing for these known determinants, there is evidence of a strong influence of transfusion styles (Chapter 9).

The causes for blood transfusion in cardiac surgery are shown in Table 10.1. An important reason for red blood cell transfusion in cardiac surgery is extracorporeal circulation since this causes a dilution of the red cell mass of the patient. For patients with high hematocrits and a large intravascular volume, this dilution rarely precipitates a need for red cell transfusion. In some patients, however, preoperative hematocrits or intravascular volume or both may be low, (such as low weight females). Under these circumstances, the extracorporeal circuit will cause a significant dilution of the red cell mass, often to a hematocrit of less than 16. Extensive resections and lack of attention to good local hemostasis will also result in excessive bleeding which may also require red cell replacement. A third reason is extracorporeal damage occurring to platelets and activation of soluble systems such as the inflammatory and fibrinolytic systems. When the patient comes off the pump and has been neutralized with protamine, this may manifest as excessive oozing. Furthermore, the use of fluids to expand the intravascular volume, such as crystalloids and/or colloids, may further dilute blood cells and coagulation factors, with a resulting dilutional coagulopathy. Attachment of platelets to a large aortic graft may result in thrombocytopenia and also contribute to a bleeding disorder, which may require treatment with blood components, either platelets, possibly plasma, or both.

Intraoperative platelet transfusion in cardiac surgery in very controversial. Prophylactic transfusions have not been shown to be effective. The rationale for the use of therapeutic platelets is the presence of unexpected, excessive bleeding (wet field) as observed by the anesthesiologist or surgeon. Since the duration and threshold for this observation prior to ordering platelets may vary from surgeon to

Clinical Transfusion Medicine, by Joseph D. Sweeney and Yvonne Rizk. © 1999 Landes Bioscience

---

Table 10.1. Reasons for blood transfusion in cardiac surgery

1. Extracorporeal circuit dilutes the red cell mass, causing anemia.

2. Excessive bleeding with dissection of the chest or graft source.

3. Long pump runs can cause platelet dysfunction, and activate the inflammatory and fibrinolytic system causing an acquired bleeding disorder.

4. Intravenous fluids and the transfusion of salvage red cells in saline will cause a dilutional coagulopathy.

5. Large aortic arch grafts will consume platelets, causing thrombocytopenia.

6. Excessive bleeding due to #3, #4, or #5 will increase the need for red cell replacement.

---

surgeon, this likely explains much of the variation in platelet use. The empiric use of plasma or even cryoprecipitate may also occur in this context, often at ineffective doses. Although use of tests of hemostasis may be helpful in guiding the transfusion of these components, in practice, the turnabout time is often too long to be of practical use. Studies using intraoperative coagulation devices with a short turnabout time have been able to reduce plasma and cryoprecipitate transfusion by measuring clotting times or fibrinogen levels. Clotting times such as the prothrombin time (PT) or activated partial thromboplastin time (aPTT) are frequently prolonged. However, a PT or aPTT ratio of 1.5 times mean in the presence of excessive bleeding is sometimes used as an indication for plasma transfusion (10-15 ml/Kg). Although hematologists often regard a fibrinogen of less than 100 mg/dl as an indication for cryoprecipitate transfusion (fibrinogen replacement), surgical services may use higher thresholds, e.g., 150 mg/dl or 200 mg/dl. Lack of agreement on the above accounts for the substantial intraoperative use, and variation in use, of blood components in cardiac surgery.

Postoperatively, excessive bleeding is manifested by an increase in the volume of chest tube drainage (> 400 ml in first two hours). This is often treated (appropriately) with red cell replacement therapy. Empirical treatment with platelets, plasma, and/or cryoprecipitate can also occur. Separating this bleeding from surgical site bleeding can be difficult with potential for over transfusion of blood components, especially platelets. Overall, institutions vary in the percentage of patients who receive platelet transfusions, from less than 5% to greater than 80%. It is likely that some patients may benefit from these platelet transfusions. However, it is also likely that a substantial number do not benefit, resulting in blood component wastage.

Modest postoperative normovolemic anemia (Hct 24-30; Hb 8-10 g/dl) is common and usually well tolerated, and the practice of *routinely* transfusing red cells to maintain the hematocrit greater than 30 (Hb > 10 g/dL) likely reflects a transfusion style.

The role of plasma and cryoprecipitate in ameliorating postoperative clinical bleeding in cardiac surgery is controversial. Mild prolongations of clotting times and modest reduction in fibrinogen are very common in postoperative cardiac patients. Administration of these products in the presence of significant clotting time prolongation time (greater than 1.5 times control) or severe reduction in fibrinogen (less than 100 mg%), is reasonable, but treatment of bleeding in the presence of borderline abnormalities may simply delay the need for surgical re-exploration.

There have been numerous approaches to reduce allogeneic blood transfusion in cardiac surgery. These are listed in Table 10.2. Predeposit autologous donation (Chapter 3) may be useful in reducing the transfusion of allogeneic red blood cells under certain circumstances. This is particularly the case if preoperative erythropoietin is used to increase the number of collections. However, there is potential danger from acute hypotension occurring during the predeposit donation in this high-risk population. In addition, this approach is cumbersome for the patient preoperatively and involves an additional expense. As such, given the low cost benefit, it is unlikely to become wide spread practice in an era of cost containment.

Desmopressin (DDAVP) was initially described in the mid-1980s as being of benefit in reducing bleeding and transfusions in patients undergoing cardiac surgery. Subsequent studies have failed to reproduce the original data with regard to the beneficial effect, and interest in the use of this drug in cardiac surgery has decreased. An agent of accepted benefit, however, is the anti-protease, aprotinin. Aprotinin is a 65 kD protein derived from bovine lung. This anti-protease has been shown in numerous studies to reduce the transfusion of red cells and other blood components. Aprotinin is known to inhibit kallikrein and, therefore, reduces the inflammatory response. Dosages are expressed in kallikrein inhibitory units (KIU). In addition, it inhibits plasmin and, therefore, reduces fibrinolytic activity. Aprotinin commonly is administered in one of two dosage regimens: 2 million KIU pre-pump; 2 million in the pump and 500,000 KIU/h as a continuous infusion post pump. Half-dose regimens have also been used and shown to be equally efficacious in reducing allogeneic transfusion. Aprotinin is a very expensive agent, and the half dose regimen is, therefore, more attractive. There has been concern in the United States with regard to postoperative graft thrombotic events,

10

Table 10.2. Approaches to reduce allogenic blood transfusion in cardiac surgery

1. Preoperative erythropoietin with, or without, predeposit autologous donation.

2. Intraoperative blood salvage.

3. Preoperative hemodilution or platelet sequestration.

4. Pharmacologic agents:

   (a) DDAVP.

   (b) Amino caproic acid or tranexamic acid.

   (c) Aprotinin.

   (d) Fibrin glue or sealant.

although studies in Europe have failed to show such an adverse effect. Other problems associated with aprotinin are the possibility of hypersensitivity and for this reason a test dose is administered initially. Aprotinin is likely to be most useful in patients undergoing extensive procedures with long pump runs or re-do procedures. Aminocaproic acid has not been as extensively formally studied as aprotinin in this patient population. Aminocaproic acid is an inhibitor of plasmin and a lower cost pharmaceutical. Empiric use has been more widespread for this reason. Topical thrombin or fibrin glue are agents which may be useful when excessive microvascular oozing occurs with difficult dissections such as re-do procedures.

Preoperative hemodilution (Chapter 3) is attractive since it supplies an autologous product with fresh platelets and blood coagulation factors to the patient. In prospective studies, however, preoperative hemodilution has been disappointing in demonstrating any decrease in the need for red cell transfusion. Platelet sequestration is a modification of preoperative hemodilution in which platelets are collected using an apheresis device, but its role in decreasing the need for blood transfusion is controversial. Lastly, intraoperative salvage of blood is common in cardiac surgery. Autologous red cells shed from the dissection fields may be aspirated into the reservoir of a salvage device, subsequently washed and reinfused. Blood from the cardiac bypass pump may be given directly intravenously. However, it is a more common practice in the United States to process this blood through the salvage machine with the red cells being returned suspended in saline. Alternatively the contents of the pump and reservoir may be ultrafiltrated; this produces a product rich in colloids with a lower total volume.

An important consideration for overall transfusion in cardiac surgery is agreement regarding thresholds at which decisions are made with regard to transfusion. These are: (1) acceptable hematocrit tolerated on the pump, (2) intraoperative platelet transfusions in suspected excessive bleeding after protamine neutralization and (3) transfusing red cells postoperatively in normovolemic patients.

Vascular surgical procedures vary greatly in potential to require the transfusion of allogeneic blood. The most important vascular surgical procedure in this regard is aortic abdominal aneurysectomy (Triple A). The procedure is typically associated with the need for a large volume transfusion of red blood cells and occasionally plasma and platelets due to the development of a dilutional coagulopathy (see Chapter 14). One of the more important aspects of managing AAA resections is the use of intraoperative blood salvage, and this can result in a dramatic reduction in allogeneic blood transfusion in these patients. The role of predeposit autologous blood and/or preoperative hemodilution in elective cases is unsettled. These patients may have compromised cardiac function and depositing blood preoperatively may potentially expose the patient to donation risk without achieving any substantial reduction in the transfusion of allogeneic blood. Other types of revascularization procedures, such as femoro-popliteal bypass or endarterectomies, are not, in general, associated with large volume transfusions. The use of intraoperative salvage has sometimes been advocated in some of these procedures, although the volume of salvaged blood tends to be minimal.

Blood Transfusion in Surgery III: Orthopedic and Urologic Surgery

Although the nature of procedures performed in orthopedic and urologic surgery differ, they have in common the potential to be often associated with blood loss, and hence the need for allogeneic transfusion. In addition, procedures in urologic and orthopedic surgery are often elective, and many such patients express interest in predeposit autologous blood donation. These similarities are shown in Table 11.1.

First, the potential to over-crossmatch allogeneic blood is prominent in both types of surgery. In orthopedic surgery, spinal, hip, and knee surgery, (particularly re-do's or bilateral procedures), and in urologic surgery, radical nephrectomies, retropubic prostatectomies, and extensive transurethral resections, there can be substantial blood loss with the subsequent need for allogeneic transfusion. On account of this potential, excessive amounts of crossmatched blood are frequently requested preoperatively for many orthopedic or urologic procedures. However, preoperatively crossmatching between 1-4 units should be acceptable, in most cases, depending on the type of procedure. For these procedures, in which blood transfusion is uncommon, a type and screen should suffice. In the event of unexpected hemorrhage, a procedure should be in place in order that blood can be dispensed expeditiously. Agreement on a maximum surgical blood ordering system (Chapter 9) is important for all of these procedures.

The practice of predeposit autologous blood (Chapter 3) increased sharply for both orthopedic and urologic elective surgical procedures throughout the 1980s, but has leveled or may be declining in the late 1990s. The elective nature of many of these procedures, the real or perceived need for allogeneic blood transfusion, and concern regarding disease transmission by blood transfusion was largely responsible for this increase. It should be noted however, that predeposit blood is over collected for these procedures, in many instances. Overall, only about 50% of all such predeposit blood is transfused perioperatively, depending on the assessment of perioperative blood loss and the tolerance of the surgeon for postoperative normovolemic anemia (Chapter 26). Opinions differ with regard to the appropriate threshold hemoglobin or hematocrit at which autologous blood should be transfused in the postoperative normovolemic patient. It has been contended that autologous blood should be transfused using the same clinical criteria as allogeneic blood. Alternatively, since autologous blood is inherently "safer" than allogeneic blood (although not without risk), it has been suggested that the threshold be different, i.e., a more liberal policy. There is no general agreement of this. It is important to appreciate that predeposit autologous blood is not completely safe

11

Clinical Transfusion Medicine, by Joseph D. Sweeney and Yvonne Rizk. © 1999 Landes Bioscience

---

Table 11.1. Similar transfusion considerations in orthopedic and urologic surgery

1. Potential to over-cross-match allogeneic blood (Chapter 9)

2. Practice of predeposit autologous blood (Chapter 3)

3. Practice of intraoperative salvage (Chapter 3)

4. Practice of acute normovolemic hemodilution (Chapter 3)

5. Limited need for plasma or platelets

6. Tolerance of postoperative normovolemic anemia

---

and reactions such as hemolysis and bacterial contamination have been reported, with potential for fatal outcome (Chapter 35).

Both types of surgery may be suitable for intraoperative salvage. Orthopedic surgery, spinal surgery and joint revisions (particularly bilateral) are appropriate indications. Intraoperative salvage should require a washing phase for the salvaged blood prior to reinfusion since particulate contaminants are common. In addition, bone chips also can sometimes clog the filter of the reservoir or the intraoperative salvage device, and aspiration should be discontinued during this phase. Importantly, aspiration should never be performed when new cement (methacrylate) has been placed. For urological surgery, a different issue arises regarding the use of intraoperative salvage in patients undergoing procedures for cancer, such as radical nephrectomies or retropubic prostatectomies. Under these circumstances, it has been suggested that blood should be reinfused using a specialized filter designed to remove leukocytes from allogeneic red cell products (R100 filter, PALL Corporation). Although, these filters have been shown by electromicroscopy to be effective in removing tumor cells, there is no data to indicate that the routine use of such filters is clinically useful, i.e., prevent metastatic spread. Avoidance of aspirating from the tumor bed itself is, however, prudent. Used appropriately, intraoperative salvage has great potential in orthopedic and urologic surgery to reduce the need for allogeneic blood transfusion.

Acute normovolemic hemodilution (Chapter 3) has been practiced on many of these patients, generally removing 2-3 units of whole blood. Although several studies in the 1980s were reported to show a reduction in allogeneic transfusion, it has been suggested that, in most instances, normovolemic dilution in itself does not result in an actual reduction in allogeneic blood transfused, but rather that increased tolerance by the surgeon for perioperative or postoperative anemia explains the observed differences. Deep hemodilution (to an immediate preoperative Hct of 20) may be useful in situations where a large blood loss (4 or more units) is likely, such as spinal fusion, but anesthesiologists are often reluctant to attempt to achieve this target dilutional hematocrit.

In both types of surgery there is a limited need for the use of plasma or platelets. The one likely exception is spinal fusion surgery in which a blood loss of 0.5-1 blood volumes or more may occur intraoperatively. Under these circumstances microvascular oozing may be encountered intraoperatively, and the use of plasma in a dose of 10-15 ml/Kg is appropriate. Procedures requiring platelet transfusions are uncommon, and this should be reserved for hemorrhage in excess of 1 blood volume (8-12 units RBC).

Last, the use of allogeneic blood in these patients will be determined to some extent by the tolerance of the surgeon for postoperative normovolemic anemia. There is a tendency to transfuse these patients, many of whom are elderly, whenever the hemoglobin falls below an arbitrary threshold of 10 g/dl. It is uncertain that these patients actually benefit from allogeneic blood transfusion in the postoperative setting at this threshold, and a threshold of 8 g/dl may be a better trigger in the absence of symptoms of hypoxemia (Chapter 26). Further studies are needed to clarify this situation.

Orthopedic surgery also presents some different clinical scenarios from urologic surgery. First, postoperative drainage and reinfusion of postoperative salvage blood continues to be practiced in orthopedic surgery. Devices are available which accompany the patient from the operating room to the postoperative area, in order to continue the collection of postoperative blood from the surgical drain. This salvage blood is unprocessed (unwashed), but routinely transfused using a filter. Although theoretically of concern because of the presence of cellular debris, this product has not been associated clinically with adverse reactions. It needs to be emphasized however, that this practice has not been shown to have an important role in reducing allogeneic exposure and it is doubtful as to whether the small amount of red cells actually harvested under these conditions effects any significant reduction in postoperative allogeneic transfusions. Second, there has been a recent interest in the treatment of patients undergoing orthopedic surgery with preoperative erythropoietin. Erythropoietin may be given in any one of a number of regimens as shown in Table 11.2. Administration may be intravenous or

11

Table 11.2. Erythropoietin in orthopedic surgery

a) To increase predeposit autologous donations

250-300 IU/Kg IV twice weekly x 2-3 weeks preoperatively

600 IU/Kg Sc weekly x 2-3 weeks preoperatively

Ferrous sulphate 200 mg daily.

b) To increase red cell mass perioperatively in anemic patients

100 IU/Kg - 300 IU/Kg SC daily x 15 doses,

10 days pre surgery and for 4 days post surgery

c) Consider in anemic patients, Jehovah's Witnesses, rare blood groups or allosensitized patients.

subcutaneous, on a weekly regimen preoperatively, or combined preoperatively and postoperatively. Published studies show that these erythropoietin treatment regimens have been associated with a reduction in the use of allogeneic red cells. Erythropoietin, when given in this situation, requires routine use of supplementary elemental oral iron. Erythropoietin will increase red cell mass and thus, increase the number of predeposited autologous blood units which can be collected; also, the increase in red cell mass will reduce the extent of postoperative normovolemic anemia thus, potentially averting the transfusion of allogeneic cells. It remains to be shown however, that, while technically feasible, this expensive intervention will translate into a patient benefit, as measured in a cost-effective analysis, given the safety of the current blood supply and the expense associated with this form of treatment. Third, European studies have recently shown a benefit of aprotinin at a dose of two million KIU in reducing acute blood loss and allogeneic transfusion in orthopedic surgery (Chapter 23). This interesting observation will require confirmation, however, in additional studies.

11

Blood Transfusion in Surgery IV: Blood Transfusion in Solid Organ Allografts

Solid organ allografts pose unique considerations regarding blood transfusion support. First, there are general considerations with regard to the blood transfusion in the context of solid organ allografts and specific consideration, related to the particular organ to be grafted.

The general considerations for solid organ allografts relate to the potential for blood transfusion to cause an undesirable outcome at the time of allografting or subsequent to allografting (Table 12.1). First, there is a need to avoid sensitization to HLA antigens, which could result in graft rejection. This is probably best achieved by the use of leukoreduced blood components, as soon as a decision has been make that the patient is a candidate for allografting. Such an approach will, however, antagonize the known beneficial effect of blood transfusion on renal allograft survival. However, with the widespread use of cyclosporine, this beneficial effect is considered less than the deleterious effect of HLA alloimmunization. The use of leukoreduced blood products, preferably by prestorage leukoreduction (Chapter 36) is, therefore, the optimal approach in these patients.

In order to prevent transfusion associated graft versus host disease (TA-GVHD), irradiation of blood products is sometimes advised in the period immediately prior, and subsequent, to allografting. The incidence of TA-GVHD (Chapter 37) associated with blood transfusion in solid organ allograft is low, and routine irradiation is not common, and represents, therefore, inappropriate practice. It should be noted that the degree of leukoreduction currently achieved with filtration is not considered adequate to prevent TA-GVHD. Potential allograft recipients who are cytomegalovirus (CMV) seronegative should receive a CMV low risk blood product (Chapter 38). By using leukoreduced blood as above, however, both sensitizations to HLA antigens and CMV risk reduction is achieved.

There are several important intraoperative considerations. Some allograft procedures fulfill the criteria for massive transfusion (Chapter 14) and many units of red cells and, on account of this plasma and platelets may be transfused. Intraoperative salvage is a frequent consideration for some of these patients because of the massive blood loss, and in liver transplantation aprotinin (Chapter 23) has been used in order to reduce blood loss. ABO incompatibilities can be a problem when the allografts contain ABO antigens to which the recipient has alloantibodies. Renal and heart allografts must be ABO compatible. Liver transplants are sometimes incompatible, on account of the supply. This will often result in diminished function or survival of the allograft.

Table 12.1. Blood transfusion considerations in solid organ allografts

I.   Preoperative/perioperative considerations:

    (a) Sensitization to HLA antigens: A concern for renal, cardiac and lung transplants.

    (b) Irradiation of blood products: Transfusion associated GVHD is very rare; not routinely indicated.

    (c) CMV risk reduced blood products: A concern for all CMV negative recipients of CMV negative allografts.

II.   Intraoperative considerations:

    (a) Potential need for massive transfusion: (Liver or double lung allografts)

    (b) ABO incompatibilities: Important for all allografts

    (c) Use of intraoperative salvage:

    (d) Use of aprotinin: (Liver transplantation)

III. Postoperative considerations:

    (a) Allogeneic leukocytes causing chimerism

    (b) Cyclosporine associated HUS requiring plasma exchange

    (c) Intravenous gammaglobulins containing red cell alloantibodies, resulting in crossmatch difficulties.

12

With regard to postoperative considerations, there is always the possibility that allogeneic leukocytes transfused with the donor organ may continue to survive, a condition called chimerism. Chimerism may complicate any organ grafting. The donor lymphocytes which survive post transplantation in an immunosuppressed environment may give rise to the production of ABO or Rhesus antibodies against the recipients red blood cells. A positive direct antiglobulin test and rarely hemolysis, may, therefore, occasionally be seen in this context. Cyclosporine itself has, in addition, been associated with hemolytic uremic syndrome, which may require treatment with plasma exchange (Chapter 40). Also the use of intravenous gammaglobulin postoperatively to attenuate graft rejection, may result in the passive transfer of red cell alloantibodies, causing difficulties with compatibility testing.

KIDNEY TRANSPLANTATION

The decision to transfuse and the choice of blood products in patients who are potential candidates for kidney transplantation has changed over the last few decades. Since the introduction of cyclosporine, current thinking is that the graft survival advantage achieved with the transfusion of allogeneic red blood cells containing a large number of leukocytes is not offset by allosensitization to HLA antigens with subsequent graft rejection. Prevention of primary HLA sensitization is important, and this can be achieved with leukoreduced blood. CMV low risk products are important for CMV seronegative recipients; CMV seropositive recipients are not known to benefit from CMV low risk blood products. Second strain CMV infection may occur in these patients, but it is considered that the second strain is acquired from the CMV seropositive allograft and not the transfused blood. If the allograft donor is CMV seropositive and the recipient CMV seronegative, there is little to be gained by the use of CMV low risk products. However, use of leukoreduced blood prior and subsequent to allografting should overcome any theoretical concerns with regards to CMV transmission in any event.

As shown in Table 12.2, red cell transfusion is uncommon perioperatively in renal transplantation.

LIVER TRANSPLANTATION

Liver transplantation presents some difficult challenges to a blood bank. Patients undergoing liver transplantation often require large amounts of all types of blood components in the perioperative period. Many of these patients have an abnormal coagulation status preoperatively and thus develop dilutional coagulopathy early with the transfusion of red cell products. In addition, after the recipient's liver has been removed, there is an anhepatic phase during which no coagulation factor synthesis occurs. During revascularization with the donor liver, an explosive fibrinolytic phase can occur. Aminocaproic acid or aprotinin have been used to attenuate bleeding from excessive fibrinolysis in this phase. In the

12

Table 12.2. Comparative median blood component use in association with solid organ allograft

Organ	Red cells	Plasma	Platelets	Cryoprecipitate
Kidney	0	0	0	0
Liver	12	13	10	0
Heart	4	5	10	0
Lung:				
Single	2	0	0	0
Double	7	2	8	0

(adapted from Tuiulzi, DJ. Transfusion Support in Solid Organ Transplantation; Eds. Reid ME, Nance SJ. Red Cell Transfusion, A Practical Guide, Humana Press Inc. Totowa, NJ)

postoperative setting, a hypercoagulable state has also been reported, which can result in thrombosis. Thus, in the earlier phase of this procedure, large numbers of red cells are required and, associated with this, the transfusion of plasma and/or platelets. If fibrinogen levels drop precipitously low, cryoprecipitate may also be transfused. Liver transplantation, when first initiated, can be associated with the transfusion of more than 100 blood components/case. As experience is gained, however, the blood transfusion requirements frequently drop by more than two thirds. The indication for transplantation may also influence the transfusion requirements; those undergoing transplantation for primary biliary cirrhosis or carcinoma use fewer blood products than those with other diagnoses, such as sclerosing cholangitis.

Patients with red cell alloantibodies often receive incompatible units of red cells early in the procedure, since they are subsequently shed during intraoperative bleeding and the more compatible red cells are transfused later in the procedure. This is in contrast to standard blood banking practice, where the most compatible blood would ordinarily be transfused first. CMV seronegative patients should receive CMV low risk products. Leukoreduction by filtration would appear optimal for these patients. As shown in Table 12.2, current blood use in liver transplantation can be considerable.

HEART TRANSPLANTATION

Heart transplantation presents many similar transfusion considerations as occur in cardiac revascularization surgery (Chapter 10). An important consideration is the need to avoid primary CMV transmission by blood transfusion in these patients perioperatively and the use of leukoreduced blood is, therefore, appropriate. Current blood use in heart transplantation is shown in Table 12.2.

LUNG TRANSPLANTS

The blood transfusion requirements in lung transplantation are dependent on whether a single or double lung transplantation is performed. Data indicates that blood transfusion requirements for double lung transplants far exceeds that of single lung transplants. Single lung transplants only require transfusion in approximately one-third of cases and median red cell use in transfused patients is only 2 units. However, over 90% of patients with double lung transplants are transfused (Table 12.2). As in the case of other solid organ allografts, use of leukoreduced cellular blood components is appropriate.

Blood Transfusion in Surgery V: General Surgery

General surgery is characterized by various procedures, many of which are infrequently associated with red cell transfusion. Because of the potential, however, to require blood transfusion in some procedures, there is often a bias to routinely request the availability of crossmatched blood or to request a type and screen prior to any procedure. For many procedures in which a blood transfusion is almost never required, there is little practical value in obtaining a blood type or antibody screen. For procedures in which there is a greater potential for transfusion (e.g., gastrectomy, low anterior resection), a type and screen is appropriate. Under these circumstances, if unexpected excessive bleeding is encountered, the transfusion of uncrossed ABO identical and Rhesus compatible red cells is acceptable. It is important to develop a list of procedures for which (a) a blood specimen is not routinely required (transfusion very rare), (b) those procedures for which a type and screen is appropriate (transfusion occasional), (c) and those procedures for which routinely crossmatching of blood is appropriate (e.g., liver resections, extensive upper abdominal resections for malignancies and colorectal surgery). This list is often called a maximum blood-ordering schedule (MBOS— Chapter 9).

An important aspect of blood transfusion practice related to general surgery concerns patients undergoing procedures in situations where the pre-procedure prothrombin time is slightly prolonged (e.g., 1-1.5 mean-control) or the platelet count is slightly reduced (50-100 x 10^9/L). Examples of such patients are those with liver dysfunction and a prolonged prothrombin time requiring a central line placement or patients with a mild degree of thrombocytopenia undergoing colonoscopy. It is common practice for some physicians to administer prophylactic plasma or platelets, respectively, in these situations. Data on the administration of plasma prophylactically for patients with *minimal hemostatic defects* shows the practice to be of no clinical benefit and, therefore, wasteful. With regard to prophylactic platelet transfusions in mild thrombocytopenia, there is also no good data to justify this practice and the actual risk of bleeding is very low. Each procedure requires consideration with regard to the degree of hemostatic compromise, the ability to visualize and control hemorrhage, if it should occur, associated abnormalities such as renal failure, and the clinical consequences of minimal excessive hemorrhage. Inexperienced operators may also increase the risk of bleeding, but prophylactic blood components will not prevent major vessel puncture. By applying these considerations, only a subpopulation of patients may be appropriate candidates for prophylactic plasma or platelets, as shown in Table 13.2.

13

Clinical Transfusion Medicine, by Joseph D. Sweeney and Yvonne Rizk. © 1999 Landes Bioscience

---

Table 13.1. Considerations regarding blood transfusion in general surgery

General:
1. Many procedures do not require red cell transfusion; tendency to over request crossmatched blood.

2. Use of components prophylactically pre-procedure such as plasma or platelets, in patients with mild coagulopathy or mild thrombocytopenia is a questionable practice.

3. Intraoperative salvage may be required in some intra-abdominal procedures.

4. Extensive intra-abdominal resections or inadvertent blood vessel section may result in massive transfusion.

Areas of Current Investigative Interest:

1. Does allogeneic blood transfusion increase postoperative infections or tumor recurrence?

2. Should leukoreduced blood be used routinely in colorectal surgery?

Table 13.2. Factors which may justify the use of prophylactic plasma or platelets prior to an invasive procedure

1. More severe hemostatic abnormality, i.e. prothrombin time
   > 1.5 mean control or platelets < 50 x 10^9/L.

2. Lack of ability to visualize or control bleeding surgically.

3. Significant clinical consequences of minimal excessive bleeding.

4. Coexistence of renal failure (creatinine > 3 mg/dl).

5. Inexperienced operator.

A similar situation may often arise with regard to patients on oral anticoagulants. Patients on therapeutic doses of anticoagulants who require elective surgical procedures should have their oral anticoagulants discontinued for approximately 48 hours. This will generally lower the international normalized ratio (INR) to 1.5. Many surgical procedures can then be performed at this INR without excessive bleeding being anticipated and immediately post procedure warfarin can be recommended. For patients requiring emergency procedures, or those with evidence of an excessive warfarin effect (INR > 3.0), plasma in a dose of 10-15 ml/kg should be given in order to prevent excessive bleeding.

In general surgery, intraoperative salvage (IAT) is common in extensive abdominal resections. Two considerations arise in this context. First, aspiration into the reservoir should be discontinued if bowel contents are in the surgical field. Second, the use of IAT in intra-abdominal malignancy resections. In general, while intraoperative salvage *can be used* for these procedures, it is best to avoid aspirating from the area of the tumor bed itself. Such blood, if salvaged, however, has been reinfused using a leukoreduction filter, primarily intended for the removal of leukocytes from red blood cell products (RC100, Pall). This filter has been shown to retain malignant cells. It is, however, of unproven clinical benefit for such patients in preventing metastatic disease and the routine use is controversial and not advised.

Extensive intra-abdominal resections such as liver resections or resections for malignant disease may occasionally result in massive transfusions. Under these circumstances, after the transfusion of 6-10 units of red cells, a dilutional coagulopathy may develop, even in patients who are hemostatically competent preoperatively. The infusion of plasma, at a dose of 10-15 ml/kg may be appropriate. If further bleeding continues, platelet transfusions may be required, particularly after 1-2 blood volumes have been transfused, depending on the initial platelet count of the patient. Early and energetic use of plasma and platelets is indicated in these patients in order to decrease total components transfused (Chapter 14).

An important area in general surgery is tolerance of postoperative normovolemic anemia. In the past, patients, particularly elderly patients, were often transfused to maintain hemoglobins over 10 g/dl (corresponding to a Hct of 30) in the postoperative state. This was considered to improve patient rehabilitation postoperatively and promote improved wound healing. With regard to the latter, no data exists showing a relationship between postoperative hematocrit and wound healing. The critical determinant of wound healing appears to be the partial pressure of oxygen [pO_2], which is independent of the hematocrit. Data for patients showing shortening of the postoperative length of stay or total hospitalization is also lacking. A study currently being conducted in postoperative elderly populations who have undergone hip replacements may help clarify the effect of postoperative normovolemic anemia on the overall course of hospital stay and rehabilitation.

There are some specific areas of current interest and controversy. First, does allogeneic blood transfusion in itself increase the risk of postoperative infections, or in the context of cancer surgery, that of tumor recurrence? Single institutional studies have shown data both supporting and rejecting such an effect. More extensive meta-analyses of these studies have failed to unequivocally show allogeneic blood transfusion to be an independent risk factor for either postoperative infections or tumor occurrence. The data linking postoperative infections to allogeneic blood transfusion is stronger, however, than that of tumor occurrence.

Further to this relationship, some randomized studies have reported that the use of leukoreduced blood results in a lower rate of postoperative infections. Other studies have failed to show such benefit, although the interpretation of each study in terms of blood product type and method of leukoreduction is complicated. In the most well conducted study in colorectal surgery, leukoreduced blood has shown a reduction in both postoperative infection rate and length of stay. This is an important area to keep under review by colorectal surgeons since it has substantial implications for optimal patient care and the overall associated costs.

13

Blood Transfusion in Surgery VI: Trauma and Massive Blood Transfusion

The blood transfusion needs of patients with severe trauma or those patients requiring massive transfusion in association with elective surgical procedures present essentially similar scenarios.

A classification of acute blood loss is shown in Table 14.1. First, there is the immediate or urgent need for red blood cells and, in some instances, other blood products. Patients presenting with acute hemorrhage with loss of less than 40% of their blood volume may tolerate fluid replacement with crystalloids, assuming a normal hemoglobin level before the acute event. A problem, however, may be estimating the loss of intravascular volume and the potential for further red blood cell loss. Ordinarily, for Class 1 and Class 2 acute trauma patients (Table 14.1), red cell transfusions are not needed, particularly in young patients who can adapt well to the acute blood loss anemia, assuming that control of hemorrhage has been, or is likely to be, achieved. If in excess of 40% blood volume loss has occurred in young patients, or less in elderly people who may have pre-existing compromised critical organ function, the urgent need for red blood cell transfusions may exist. Two difficulties arise in this setting. (1) The circumstances may not allow the collection of a sample for routine compatibility testing (Chapter 7). Transfusion of blood group O red cells to these individuals is an appropriate early measure. Rhesus negative units should be used, if possible, and in all situations for females of child bearing age, arbitrarily under the age of 50 years. (2) If more time allows, a blood sample can be collected. Unfortunately the normal identification mechanisms for insuring sample integrity may not be followed appropriately because of pressures in dealing with patient resuscitation. An inappropriately labeled specimen or a misidentified specimen is then received in the blood bank, resulting in frustration on the part of the emergency room and blood bank personnel. For inappropriately labeled specimens, the continued release of group O blood remains necessary. Mislabeled specimens are particularly dangerous in this setting as the stage is set for an acute hemolytic reaction (Chapter 32). It is essential to collect and label the specimen correctly at the point of sample collection and the phlebotomist must sign (and date) the specimen. A specimen collected into an unlabeled tube, which is removed from the point of collection and labeled elsewhere, is dangerous. In summary, if time precludes adherence to correct labeling protocol, it is better to continue to transfuse group O (uncrossmatched) blood.

A second problem in the emergency room setting is the logistics of red cell availability. An effective mechanism to ensure that red cells can be rapidly delivered to the emergency room is imperative. The physical location of the blood bank in close proximity to the emergency room is helpful, and this is often the

14

Table 14.1. Classification of acute blood loss

	Class I	Class II	Class III	Class IV
Blood Loss (ml)	< 750	750-1500	1500-2000	> 2000
% Blood Volume	< 15%	15-30%	30-40%	> 40%
Clinical:				
Pulse rate (min)	< 100	> 100	> 120	> 140
Blood Pressure	Normal	↓	↓↓	↓↓↓
Respirator rate (min)	< 20	20-30	30-40	> 35
Fluids	Crystalloids only	Crystalloids; Possible Blood Transfusion	Crystalloids; Probable Blood Transfusion	Blood Transfusion and Crystalloids

Adapted from the Advanced Trauma Life Support Subcommittee of the American College of Surgeons

Table 14.2. Considerations regarding massive blood transfusion

- The immediate or urgent need for red blood cells and other products may preclude routine compatibility testing.

- Error in specimen identification may occur.

- Rapid transfusion of red blood cells may cause hypothermia.

    (a) Blood warmers

    (b) Warm saline mixing with blood

- Complications of large volume transfusion over a short (< 24 hours) time period.

    (a) Metabolic

    (b) Dilutional

- Follow-up cohort to examine risk of disease transmission by blood transfusion.

case in many hospitals with Level I Trauma Units. In the absence of this, there may be a need to have group O Rhesus negative blood available in a refrigerator in the emergency department. A "trauma pack" (two Group O, Rhesus negative/two Group O, Rhesus positive red cells) constitutes a reasonable stock. The desirability of this approach needs to be balanced, however, by a reliable and consistent mechanism to ensure adequate identification of recipients in order to have complete records of disposition for any blood transfused and to avoid unnecessary transfusions, since ease of availability may promote earlier use in situations where crystalloids may be adequate.

A further problem is the need to transfuse red blood cells rapidly in a moribund, hypotensive patient. Red cells are stored between 1-6°C, and rapid transfusion of large volumes may cause hypothermia. This occurs particularly at rates of infusions greater than 100 ml/minute, equivalent to the transfusion of a unit in 2-3 minutes. The use of a blood warmer is helpful in this setting, though it must be ascertained that the blood warmer is effective at rapid transfusion rates, as some blood warmers are only effective at warming blood at slower infusion rates. Blood warmers typically have a maximum temperature which is less than 42°C, and the blood is typically warmed to 37°C. Other approaches to warm blood have been used, such as the rapid addition of prewarmed saline (68°C) to red cells prior to infusion (red cell admixture). This practice is of some concern, since red cell hemolysis may result, with substantial increases in potassium, causing metabolic complications (see below). It is best avoided, unless considerable in-house experience exists with the approach.

A massive transfusion is arbitrarily defined as the transfusion of more than one blood volume (BV) over a 24 hour period. Calculating blood volumes as an absolute number of units transfused is inappropriate, particularly for low weight (female) adults. Thus, one BV transfusion could result from the transfusion of as few as six units of allogeneic cells in such an adult. Complications associated with massive transfusions are best divided into metabolic and dilutional (Chapter 32).

The important immediate metabolic problems associated with blood transfusion relate to the potential for high concentrations of potassium to cause hyperkalemia and/or rapid citrate infusion to cause hypocalcemia. The hyperkalemic problem arises since the extracellular concentration of K^+ increases progressively during red blood cell storage from approximately 4 mEq/L at the time of collection to approximately 40 mEq/L at the end of the maximum storage period (42 days). If irradiated blood is used (uncommon in the context of acute trauma or massive transfusion), then the K^+ concentration in the red cell product could exceed 70 mEq/L. Although these concentrations of K^+ are very high, the absolute amount of potassium transfused per unit is modest. Transfusing six units of blood at 42 days of storage over a one hour period (35 ml/blood /min for 60 minutes) is equivalent to transfusing a total of approximately 40 mEq K^+/h. This amount of potassium given to a trauma patient should be well tolerated and would, at most, increase the K^+ concentration no more than 2 mEq/L. A decrease in K^+ from hypercatabolism will, in addition, antagonize any increase. Thus, in practice, hyperkalemia should not be a problem, except in certain situations, such as in patients with renal failure.

The second metabolic problem, which may occur, is due to the large amounts of citric acid. This does not occur commonly with red cells, since most red blood cells transfused are stored in crystalloid solution which contain little of the original citrated plasma. Citric toxicity occurs in massive transfusion due to the rapid transfusion of large volumes of plasma (> 20 ml /Kg) or platelets, which are stored in citrated plasma. Severe hypocalcemia can cause convulsions and hypotension. In most situations, the use of calcium is unnecessary and maintaining a vigilance for symptoms of hypocalcemia is reasonable. Calcium chloride is preferred, since

14

ionized calcium is more readily available. The ability to metabolize citric acid is dependent on liver size and function, and citrate toxicity is more likely to occur in low weight females. Citrate can be useful in that it will increase the bicarbonate levels in plasma and promote a metabolic alkalosis. In practice this will antagonize the metabolic acidosis associated with hypoxemia, which is the more important acid-base disturbance in these patients, and also help reduce K⁺ in plasma. Dilutional coagulopathy is a more common problem. In the older literature, it was often considered that platelet transfusions were appropriate after one BV transfusion. This data however, came from an era (pre-1982) when red cells were stored as CPD or CPP-A1 red cells (Chapter 2), i.e., in anticoagulated plasma. At the present time, red cells are mostly stored in additive solutions, and the transfusion of large volumes of red cells will very rapidly cause a dilution and a reduction in blood clotting factors. As little as 0.5 BV transfusion, (arbitrarily 5-6 units of red cells) may result in clinically evident microvascular oozing. This is best managed with the initial infusion of fresh frozen plasma at a dose of 10-15 ml/Kg. This replaces the deficiency of clotting factors, particularly factor V and fibrinogen, as these factors are largely distributed intravascularly. After further transfusion of red cells (total 1-2 BV), there is the potential for thrombocytopenia ($< 50 \times 10^9$/L) to be an important complication depending on the initial (pre-transfusion) platelet count. Platelet transfusion (1 U/10 Kg) may then be appropriate, particularly if the platelet count is 50×10^9/L or less, and microvascular oozing is present.

The guideline for the transfusion of either plasma or platelets should be the patient's estimated intravascular blood volume. The dosing of plasma transfused should be in ml/Kg, not "units FFP (1 unit FFP = 200-220 ml). Similarly, dosing of platelets is important since lower weight individuals will respond very satisfactorily to smaller doses of platelets, for example, 4 or 6 units may be acceptable. The relationship between blood volume loss and the degree of dilutional effects on intravascular cells or high molecular weight clotting factors is shown in Table 14.3.

In the past, patients who received massive transfusion were followed prospectively since they constituted a cohort exposed to large numbers of blood components and were useful in estimating the risk of disease transmission by blood transfusion. However, the risk of disease transmission is now so low that this approach is unlikely to yield useful information.

14

Table 14.3. Dilution of platelets and high molecular weight intravascular clotting factors (factor V) after large volume transfusions over a short period (< 24 hours)

Amount of blood transfusion (either allogeneic red blood cells in additive solution or salvaged autologous red cells)	Residual concentration of platelets or high molecular weight clotting factors (factor V, fibrinogen).
1 BV	33%
2 BV	25%
3 BV	12%
4 BV	4%

BV = Blood volume

Blood Transfusion in Medicine I: Cancer

The transfusion supportive care of patients with cancer is responsible for a large proportion of blood transfused within developed countries such as the United States, Europe and Japan. From the perspective of blood transfusion, it is useful to group patients with cancer into three different categories. First, hematologic malignancies in adults, which although comprising only 10% of all cancers in this population, account for much of the blood product use, especially platelets. Second, adult non-hematological malignancies, which, are mostly treated with local forms of treatments, such as surgery or radiation therapy. Although, chemotherapy may be used, cytopenias resulting from chemotherapy which require transfusion support are uncommon, outside of the context of lung, breast or ovarian carcinomas. Third, pediatric malignancies. Approximately 50% of pediatric malignancies are hematological malignancies; however, solid tumors which occur in children, such as neuroblastomas, are more likely to result in chemotherapy-related cytopenias and transfusion support more closely resembles that of adult patients with hematologic malignancies.

HEMATOLOGICAL MALIGNANCIES IN ADULTS

The major hematological malignancies requiring blood transfusions are the acute leukemias, advanced stage lymphomas, myelomas, myeloproliferative and myelodysplastic disorders. Early stage lymphomas and many of the chronic leukemia in early stages are uncommonly associated with blood transfusion. There are several important blood transfusion considerations in adults with hematological malignancies.

THE USE OF LEUKOREDUCED BLOOD

In general, it is preferable to use leukoreduced blood for all patients with hematological malignancies. The rationale is to prevent primary alloimmunization to HLA antigens, since many of these patients ultimately will require platelet transfusions. Also, many of these patients will require multiple red cell transfusions, and avoidance of transfusion reactions is always desirable in multiply transfused patients (Chapter 32). The use of blood leukoreduced by filtration has been shown to be cost-effective in acute leukemia in that the reduction in sensitization to HLA antigens reduces the subsequent need for expensive HLA selected platelet products. An overall policy therefore to use leukoreduced blood in these patients is appropriate.

Clinical Transfusion Medicine, by Joseph D. Sweeney and Yvonne Rizk. © 1999 Landes Bioscience

CYTOMEGALOVIRUS LOW RISK PRODUCTS

Some patients with hematological malignancies who are CMV seronegative, may be appropriate candidates for CMV low risk products, if they are potential or actual candidates for allogeneic bone marrow transplantation. These patients should always receive CMV low risk products. In the past, the only acceptable CMV low risk product was a component from a donation which was serologically negative for CMV. However, leukoreduction by a method which prevents alloimmunization (less than 5×10^6 residual white cells), is considered essentially equivalent to CMV seronegative blood. Thus, a policy to use leukoreduced blood in patients with hematologic malignancies will also achieve the objective of preventing primary CMV transmission.

IRRADIATED PRODUCTS

Irradiated products constitute another controversy in patients with hematologic malignancies. The most important disease in this category is Hodgkin's disease. Although more cases of transfusion associated-graft-versus-host disease (TAGVHD) have been reported with acute leukemias than Hodgkin's disease, the known cellular immune defect of Hodgkin's disease has received considerable prominence as predisposing to TAGVHD (Chapter 37). A policy to give irradiated products to all patients with Hodgkin's disease is appropriate. Some institutions provide irradiated products for all patients with hematologic malignancies. This is, however, neither a widespread nor an accepted practice and there is little data to indicate that the routine use of irradiated products in patients with acute leukemias, lymphomas other than Hodgkin's disease, or plasma cell dyscrasias is of benefit. However, the use of irradiated blood is appropriate at certain times in patients with acute leukemia or non-Hodgkin's lymphomas who are candidates for bone marrow transplantation (Chapter 16).

Last, there is the question of the role of erythropoietin in patients with hematologic malignancies. This relates particularly to patients with myelodysplastic disorders. Other patients with hematologic malignancies have defects in late stem cell progenitors due to chemotherapy or tumor crowding and, as such, the ability of erythropoietin to improve the anemia is more limited.

Platelet transfusions in patients with hematologic malignancies should be leukoreduced, prestorage, if at all possible (Chapter 28). This is useful in reducing bedside transfusion reactions such as fever and chills (Chapter 32). Red cells are preferably filtered prestorage, but bedside filtration is also effective in reducing reactions and preventing primary alloimmunization to HLA antigens.

SOLID TUMORS IN ADULTS

Solid tumors represent approximately 90% of all adult tumors, and blood transfusion issues are very different. First, many of these patients are managed by local forms of therapy, such as surgery and radiation therapy: chemotherapy is supplementary or adjuvant to management. Exceptions to this rule, are small cell carcinoma of the lung and testicular tumors. The blood products most commonly used are red cell products which are transfused perioperatively due to bleeding or on account of radiation induced myelosuppression. There is an ongoing controversy as to whether allogeneic blood transfusion is an independent risk factor in increasing tumor recurrence post surgery (Chapter 13). At this time, therefore, it would appear desirable to avoid allogeneic blood transfusion, if at all possible. A related controversy is the role for leukoreduced blood in preventing tumor recurrence. Since it is unsettled whether allogeneic blood transfusion is a determinant of tumor recurrence, the potential role of leukoreduction in this context is unclear.

Second, there may be a role for blood transfusion to increase the radiosensitivity of tumors. Tumors may be more responsive to radiation therapy in the presence of well oxygenated blood, which acts as a radiosensitizer. When these patients are transfused to higher hematocrits (> 35), more oxygen may be off-loaded at target tissues, giving rise to an enhanced radiation effect. This effect is also being investigated with blood substitutes, such as the animal derived hemoglobins and the perfluorocarbons.

Third, extensive surgery in solid tumors can result in the need for massive transfusion. Most surgical procedures for patients with cancer are associated with modest use of blood products (less than 4 units of RBC). However, more extensive cancer surgery, particularly in patients who are anemic preoperatively, will be associated with large volume red cell transfusions, and the subsequent need to transfuse plasma, and, possibly, platelets (Chapter 14).

Outside of the context of massive transfusion, use of blood components other than red cells is not common in solid tumors. Chemotherapy associated cytopenias do occur, however, in ovarian carcinoma, small cell carcinoma of the lung and breast cancer, and platelet transfusion may be required. It is uncertain whether these populations of recipients benefit from leukoreduced blood products, although patients who require treatment with multiple courses of chemotherapy with associated thrombocytopenia will likely require platelet transfusion and, therefore, benefit from leukoreduced blood products.

PEDIATRIC MALIGNANCIES

Pediatric malignancies differ from adult malignancies in that a larger percentage of the tumors are hematologic tumors and chemotherapy is often the primary form of therapy. For this reason, in general, transfusion support of patients with pediatric malignancies is closer to the treatment of adult hematologic malignancies and the same issues and controversies exist with regard to the use of leukoreduced blood, CMV low risk products and irradiated products. It is prudent to treat all of these patients with leukoreduced (filtered) blood from the outset. Prevention of transfusion transmitted CMV arises frequently because most of these younger patients are CMV seronegative. This is probably best managed with the universal use of leukoreduced blood. A particular area of controversy is the common practice to irradiate all cellular blood products for patients with pediatric malignancies. Surveys have demonstrated that many centers have a strong

15

preference for the universal use of irradiated blood products in patients with pediatric malignancies. Irradiated blood products increase cost, alter the logistics of blood product supply and may constitute a wasteful practice in some cases. However, for many patients with pediatric malignancies, there is concern regarding the immune status, particularly those patients less than one year old, for example, with neuroblastomas or in situations where hereditary *disorders of the cellular immune system* may coexist. A careful review on a case-by-case basis of the real need for irradiated products with restricted use to more well defined situations (such as Hodgkin's disease, bone marrow transplant recipients and lymphomas associated with immune defects), would appear reasonable. In practice, however, universal irradiation is often employed because of concerns that an individual patient with an appropriate indication could receive a non-irradiated product in error, resulting in a catastrophic outcome (Chapter 37).

Table 15.1. *Considerations regarding blood transfusion in cancer*

I.   Hematologic Malignancies:

    (a)   Leukoreduced cellular blood products advisable for all patients.

    (b)   CMV risk reduced products for some CMV seronegative recipients.

    (c)   Irradiated products for specific indications.

    (d)   Role of erythropoietin in myelodysplastic states.

II. Non-Hematologic Malignancies:

    (a)   Does allogeneic blood transfusion increase tumor recurrence?

    (b)   Is there a role for transfusion to radiosensitive tumors?

    (c)   Extensive surgery associated with potential for massive transfusion.

III. Pediatric Malignancies:

    (a)   Preference for the use of leukoreduced, CMV risk reduced and irradiated products for all recipients.

15

Blood Transfusion in Medicine II: Bone Marrow Transplantation

Blood transfusion support is essential to the successful outcome of bone marrow transplantation, and the presence of a bone marrow transplant unit in an institution will likely account for a large percentage of total platelet usage. Bone marrow transplantation is more correctly termed "stem cell transplantation", since there are several sources of hematopoietic progenitor cells (HPC). Traditionally, HPC have been collected by large volume aspiration of bone marrow under anesthesia (1-2 liters). In the last fifteen years, there has been increasing interest in collecting HPC from peripheral blood, initially for autologous, and more recently, for allogeneic transplantation. Newer sources of stem cells are umbilical cord blood; fetal hepatocytes may prove useful in the future. At this time, much interest is focusing on peripheral blood and umbilical cord blood as sources of HPC.

Stem cell transplantation is best divided into the autologous and allogeneic, since transfusion requirements and considerations differ.

AUTOLOGOUS STEM CELL TRANSPLANTS

The important transfusion considerations are shown in Table 16.1. First, the use of leukoreduced blood is recommended. The primary purpose is to prevent HLA alloimmunization and, thus, avert problems with refractoriness to platelet transfusions due to HLA alloantibodies. This will also suffice as a means of preventing the primary transmission of cytomegalovirus (CMV) in patients who are CMV seronegative. Second, use of irradiated blood products. It is most important that the patient receive irradiated blood in the period (2 weeks) immediately prior to any stem cell collection, whether by apheresis techniques or from bone marrow aspiration. This is to prevent the transfused allogeneic leukocytes in donor blood being harvested, cryopreserved and subsequently causing transfusion-associated graft versus host disease after transplantation of the stem cell product. Irradiated blood should be routine once conditioning has begun and continues until approximately 3-6 months after engraftment. Third, autologous stem cell transplants are associated with considerable use of red blood cells and platelets. The judicious use of cytokines, such as G-CSF, will facilitate white cell recovery but platelet recovery tends to lag behind. Platelet support generally continues up to 15 days after transplantation, (either daily or alternate day, depending on dosage). The use of cytokines, such as thrombopoietin has been disappointing to date in reducing this requirement.

16

Clinical Transfusion Medicine, by Joseph D. Sweeney and Yvonne Rizk. © 1999 Landes Bioscience

Table 16.1. Consider regarding blood transfusion in autologous stem cell transplants

1. Use of leukoreduced blood to prevent HLA alloimmunization and CMV transmission.

2. Use of irradiated cellular blood products prior to stem cell collection and from conditioning until 3-6 months after engraftment.

ALLOGENEIC STEM CELL TRANSPLANTS

Allogeneic stem cell transplants present some different problems in regard to blood transfusion support. First, there are considerations regarding the stem cell donor and use of family members as directed donors for blood products. Family members are best to avoid as blood donors, since the use of family members may give rise to allosensitization to minor histocompatibility antigen and hence subsequent graft rejection. Second, the healthy donor may have a large volume aspirate if bone marrow is used as the source of HPC. Predeposit of autologous blood is a common practice but occasionally allogeneic blood may be transfused. It needs to be emphasized that healthy donors are capable of withstanding considerable amounts of acute blood loss anemia, (Chapter 26) and thus the decision to transfuse allogeneic blood should be made conservatively.

Recipient considerations have similarities and differences from autologous transplant. First, leukoreduced blood is routinely recommended to prevent HLA alloimmunization, and recent studies have indicated that the use of leukoreduced blood by filtration is adequate to prevent CMV infection in CMV seronegative recipients of CMV seronegative allogeneic transplants. CMV disease is a matter of great concern, however, on the part of transplant physicians, and there is a strong preference for the use of CMV seronegative blood for CMV seronegative recipients. The use of CMV seronegative blood for CMV seropositive recipients is based on the concept that a second strain CMV infection may occur. This has never been documented by blood transfusion, although it is known to occur in solid organ allografts (Chapter 12) where the recipient is CMV seropositive *and* the organ donor is CMV seropositive. Third, irradiated cellular blood products should be used prior to commencement of conditioning until at least two years (on indefinitely) after engraftment. The timing of the use of irradiated blood in the context of allogeneic transplants differs, therefore, from the use of irradiated blood in autologous transplants. The risk for transfusion associated graft versus host disease is greatest at the time of immunosuppression, which follows conditioning, and the period of marrow hypoplasia in the early phases after engraftment. Although it is unclear for how long patients should receive irradiated blood, the indefinite use of irradiated blood post transplantation may be wise, particularly if there is evidence of ongoing need for intense immunosuppression. Fourth, the use of red cells and platelets is far greater in allogeneic than in autologous transplants (Table 16.2). It is not uncommon for patients to require red cell transfusions and sometimes platelet transfusions for several months after the transplant

16

Table 16.2. Comparison of allogeneic blood transfusion requirements in autologous and allogeneic stem cell transplantation

	Red Blood Cells (units)	Platelets (Transfusion Episodes)
Autologous	10	11
Allogeneic	24	29

(Data given are the mean; within each group, a large inter subject variation exists)

Table 16.3. Blood transfusion considerations in allogeneic stem cell transplants

I. Donor considerations

   (a) Use of family members as blood donors is inadvisable.

   (b) Possible need to transfuse red cells to stem cell donor; encourage predeposit donation.

II. Recipient considerations

   (a) Use of leukoreduced blood to prevent HLA alloimmunization

   (b) Use of CMV seronegative blood, if recipient is CMV negative

   (c) Use of irradiated cellular blood products at the commencement of conditioning until at least 2 years after engraftment.

   (d) Use of RBC/platelets is 2-3 times greater than autologous transplants

   (e) Group O RBC preferred in most instances

   (f) ABO incompatibility may cause acute or delayed effects

   (g) Antibodies to minor blood antigen systems may cause difficulty in red cell compatibility testing.

procedure. Group O red blood cells are preferably used in allogeneic transplants regardless of the ABO type of the recipient or donor. This is to avoid problems with incompatibilities, which may give rise to hemolytic events in vivo. ABO incompatibilities may cause significant problems in allogeneic transplants. In the first instance, the recipient may have ABO alloantibodies, which are incompatible with the donor red cells. Bone marrow derived stem cell products are heavily contaminated with red cells, to the extent that the equivalent of a single unit of blood may be administered during the reinfusion of a stored stem cell product. In order

16

to circumvent this problem, various approaches have been tried such as deplet the recipient of ABO antibodies by plasma exchange or immune-absorption, these are difficult and often ineffective. A preferred approach is to deplete marrow stem cell product of red blood cells by the sedimentation of red c prior to reinfusion. Despite these maneuvers however, both acute and sometin delayed hemolysis due to ABO system antibodies may occur. In addition to ac or delayed effects due to alloantibodies in the ABO system, antibodies to mii blood group antigens, such as Rhesus, may also be observed in the period p transplantation. This can be due to passive transfer of antibodies by the use intravenous gammaglobulin, but microchimerism has also been described in wh a residual recipient population of immunocytes produce antibodies against nor red cell antigens (Chapter 12). The ABO type of the red cell will change to t of the donor much later, generally 45-65 days, after the transplant. ABO inco patibilities are known to delay engraftment, as manifested by a low reticuloc count and an associated prolonged need for red cell transfusion support.

Blood Transfusion in Medicine III: Hereditary Anemias

Blood transfusion support in patients with hereditary anemias differs in that) many of these patients receive their first transfusion in early life, and since ng term chronic red cell transfusion is often employed iron accumulation by olescence presents a clinical problem (2) although the need for transfusion is rtly related to the need to increase oxygen delivery, an important aspect is the ppression of erythropoiesis by elevating the hematocrit, which has the effect of eventing developmental skeletal malformations and (3) use of blood products her than red blood cells is unusual.

The major hereditary anemias which require red cell support are shown in ble 17.1, but most red cells will be transfused in the treatment of patients with kle cell syndromes and the thalassemias.

SICKLE CELL SYNDROMES

The blood transfusion support of patients with sickle cell syndromes includes tients with hemoglobin SS disease, patients with hemoglobin SC disease, and moglobin Sβ° thalassemia.

The first consideration in the transfusion of patients with sickle cell syndromes to define a desired post-transfusion hemoglobin (target). Without transfusion, ese patients will have a hemoglobin in the range of 5-8 g/dl, and transfusing to moglobins in excess of 12 g/dl is probably inadvisable because of concerns re- rding high blood viscosity. In practice, this is usually a consideration only when nsfusing patients in the context of preparation for surgical procedures. The cond consideration with sickle cell syndromes is that these patients have a high opensity to develop antibodies to transfused allogeneic red blood cells. This curs in 25-45% of chronically transfused patients. The reason for this oimmunization is twofold. (a) Patients with sickle cell syndromes are mostly rican/American. The blood donor population in the U.S. is mostly European- nerican. European-American antigens within the Rhesus and minor blood group stems differ from African-Americans. The most important differences reside in o antigens within the Rhesus system (designated C and E) and in that nearly all rican-Americans (98%) lack an antigen within a blood system called Kell (des- nated K-1, but usually called Kell). In addition, African-Americans usually lack o common antigens within a system called Duffy (designated Fyᵃ and Fyᵇ), and wer express an important antigen within a system called Kidd (designated Jkᵇ).

17

inical Transfusion Medicine, by Joseph D. Sweeney and Yvonne Rizk. © 1999 Landes Bioscience

---

Table 17.1. Hereditary anemias in which red cell transfusion is common

  I. Sickle cell syndromes

 II. Thalassemias

III. Miscellaneous group:

    (a) RBC membrane defects—e.g., spherocytosis

    (b) RBC enzymopathies—e.g., pyruvate kinase deficiency

    (c) Congenital dyserythropoietic anemias

    (d) Hereditary aplastic anemias

Table 17.2. Considerations regarding transfusion in the sickle cell syndromes

1. What is the optimal (desired) hematocrit?

2. Antibodies to allogeneic red blood cells (25-45%) complicate blood availability.

3. Use of leukoreduced red blood cells to prevent nonhemolytic febrile reactions.

4. Preoperative transfusion and/or exchange transfusion.

5. Red cell exchange in acute chest syndrome, priapism or stroke.

6. Miscellaneous

    (a) Splenic sequestration

    (b) Hyperhemolysis

    (c) Transient aplasia

    (d) Iron overload

In practice, therefore, antibodies in sickle cell patients develop to the antigens (
E, Kell and, less commonly, to Fy^b, Fy^a and Jk^b. (b) In addition to exposure to the
antigens however, it is considered that there may be a genetic predisposition
these patients to form alloantibodies. When multiple red cell antibodies develo
it can present technical difficulties and delays in providing compatible red blo
cells. In an attempt to circumvent this, a program known as "Phenotyped Match(
Blood" has been attempted in some institutions. In this program, suitable dono

are identified (mostly of African/American origin) who have similar antigen profiles to the sickle cell population. Transfusing this type of donor blood has been shown to greatly reduce the risk of alloimmunization, as would be expected. Another reason to reduce the occurrence of alloimmunization is that multi-transfused sickle patients sometimes develop autoantibodies. While these autoantibodies rarely cause hemolysis in vivo, they complicate compatibility testing for these patients. Third, patients receiving multiple red cell transfusions as a general rule (and sickle patients are no different), should receive leukoreduced red blood cells: (a) to reduce the discomfort of acute reactions to red cell transfusions (Chapter 32) and (b) to prevent any confusion with hemolysis, since hemolysis is always suspected when a reaction occurs in these patients on account of the problem of alloimmunization. Fourth, preoperative transfusions and/or exchange transfusions are sometimes used in sickle cell disease prior to surgical procedures. This is a common practice in some centers in the United States but, in other countries where sickle cell disease is common, such as Nigeria, clinical experience suggests that this is not necessary. If careful attention is paid to intraoperative avoidance of acidosis, hypoxemia and hypovolemia, surgery can be performed even with low hematocrits. Transfusion is best reserved for perioperative bleeding. It is likely that the young age of these patients and the chronic nature of the anemia makes this approach feasible. Fifth, all blood products transfused to these patients should be tested for the sickle cell trait and should test negative. Sixth, red cell exchange (Chapter 40) is used in certain clinical situations, particularly in the treatment of the acute chest syndrome (and possibly priapism or stroke). Acute chest syndrome is a life threatening complication associated with sickle cell disease. Red cell exchange is an apheresis procedure (Chapter 40) in which the red cells of the patient are exchanged with compatible donor red cells. A target reduction of the hemoglobin S concentration to less than 30% is usual. Red cell exchange has also been used in other situations, such as priapism, and in the treatment of cerebrovascular events. Seventh, there are a series of miscellaneous clinical situations, which impact the transfusion support of these patients. In *splenic sequestration* and *transient aplasia*, a life-threatening anemia can arise in which a very rapid reduction in hemoglobin to values as low as 2-3 g% occurs, requiring aggressive transfusion support. Hyperhemolysis is an unusual complication in which the patient does not respond to transfused allogeneic blood with the expected increase in hemoglobin. This is perplexing both to the treating physician and blood bank, since the hemoglobin post transfusion is sometimes even lower than the hemoglobin pretransfusion! Transfusion is best temporarily discontinued. The pathophysiology of the hyperhemolysis is unknown, but may be related to complement activation on the surface of the transfused cells. Eight, iron overload is a consideration in multitransfused sickle cell syndrome patients and needs consideration as the number of transfusions accumulate (see discussion on Iron in Thalassemia).

17

Table 17.3. *Considerations regarding transfusion in the thalassemias*

1. Hypertransfusion or supertransfusion?

2. Alloantibodies to red blood cells (20-40%) and the role of extended phenotyping.

3. Use of leukoreduced red blood cells to prevent febrile nonhemolytic reactions.

4. Use of neocyte (younger red cell) transfusions.

5. Development of iron overload.

THALASSEMIC STATES

Some transfusion issues are common to sickle cell syndromes and thalassemia. First, although alloimmunization does occur in a significant proportion of these patients (5-25%), thalassemics, however, tend to have the same blood group antigen distribution as the donor population (i.e., European American), and hence red cell allosensitization to uncommon red cell antigens is less of a practical problem. Patients with thalassemia often have an extended red cell phenotype performed before starting a transfusion protocol and this facilitates the subsequent identification of alloantibodies. Second, the question of transfusion threshold and target differ. In some centers, hypertransfusion (arbitrarily to maintain a hemoglobin greater than 10 g/dL) is widely used to prevent the skeletal deformations associated with thalassemia. This transfusion protocol commences before the age of three. Supertransfusion has also been used, which strives to increase the hemoglobin beyond 13 g/dl but, this has not shown to be of more benefit than the standard hypertransfusion protocols. Therefore, transfusion of these patients at a threshold pretransfusion Hb of 9-10 g/dl is generally employed. Third, leukoreduction should be used in all patients for the same reasons as the sickle cell population and leukoreduction by filtration has been shown to reduce the occurrence of nonhemolytic febrile transfusion reactions in this population. Fourth, there has been interest in the use of a subpopulation of donor red cells, called neocytes, for thalassemics. These are the reticulocyte rich fraction of donor blood which are less dense than the more mature red blood cells. This difference in density allows neocytes to be separated by centrifugation and "neocyte red blood cell products" manufactured. Use of these neocyte transfusions has been shown to increase the interval between transfusion therapy, which in thalassemics is ordinarily about once every 2-4 weeks. The complexities, however, with the production of neocytes probably preclude the widespread use of this technology. Last, iron overload is a major problem in patients with thalassemia on hypertransfusion protocols. One unit of blood contains approximately 250 mg of iron and, therefore, considerable iron accumulation in the heart occurs with time, giving rise to a congestive cardiomyopathy. The management of this problem is beyond the scope of this handbook, but in general, the monitoring of serum ferritin is important ith commencement of iron chelation therapy when ferritin exceeds 1500 ng/ml. Although there is interest in oral chelators of iron therapy, the only treatment known to be effective at this time is subcutaneous infusion of desferrioxamine.

Although the sickle cell syndromes and thalassemic states account for most of the red cells transfused in the support of the hereditary anemias, there are a miscellaneous group of hereditary diseases which require red cell transfusion. These include disorders such as hereditary spherocytosis, in which splenectomy is the more definitive form of treatment. In hereditary spherocytosis, splenectomy is avoided until the age of six and transfusions may be required to suppress erythropoiesis and achieve a hemoglobin which allows for normal skeletal development prior to this. Red cell enzymopathies, particularly pyruvate kinase deficiency, may also require transfusion, prior again to splenectomy. The threshold for transfusion in pyruvate kinase deficiency is less, however, since the level of 2, 3 diphosphoglyceric acid is higher in these cells, allowing better oxygen off-loading. Uncommon entities such as a congenital dyserythropoietic anemias and some hereditary aplastic anemias may also be associated with the need for chronic transfusion. In this miscellaneous group, extended phenotyping is often performed, as in the thalassemias, to facilitate antibody identification in the event of the occurrence of alloimmunization.

Blood Transfusion in Medicine: IV: Renal Disease

Patients with renal disease who require blood transfusion impact a blood bank in several different ways. First, there is the question of appropriate transfusion support for patients who are potential renal transplant candidates. In recent decades there has been considerable change in the approach towards the type of red cell product transfused to these patients. In the 1950s, blood transfusions were routinely administered in order to improve symptomatic anemia. This practice was discouraged in the early 1960s, because of concerns regarding allosensitization to HLA antigens, which would result in subsequent renal allograft rejection. In the later 1960s, it became evident that patients who had been transfused pretransplant paradoxically showed an improved graft survival!! This was attributed to an immunomodulatory effect of blood transfusion. For this reason, blood transfusions (generally > 3 units) using standard nonleukoreduced red cells were intentionally given pretransplantation in this patient population, throughout the 1980s. With the availability of cyclosporine A, however, current thinking is that the risk of alloimmunization to HLA antigens from white cell rich red cell transfusions exceeds the benefit of the immunomodulation, such that "active transfusion" is no longer considered appropriate (Chapter 12). Therefore, the question arises as to what type of red cell product would be most appropriate, if needed pretransplantation.

As shown in Table 18.1, the suggested product is a leukoreduced (filtered) red blood cell, preferably a prestorage leukoreduced product. This blood product will prevent primary HLA alloimmunization thus reducing allograft rejection. In addition, this product will also constitute effective prophylaxis for CMV transmission. Historically, washed red cells were requested on the basis that this will result in a reduction of HLA sensitization. Washing however, is very ineffective in removing allogeneic leukocytes (only about 70% are removed) and the preferred approach is the use of a filtered product, which removes 99.99% of leukocytes (Chapter 36).

Patients on chronic dialysis who are not transplant candidates present a different situation. In the past, transfusion of dialysis patients accounted for 3-5% of all red cells transfused in the U.S. Erythropoietin has greatly reduced the number of transfusions in this population. Despite this, there are some patients on dialysis who still require transfusion. Multiple transfusion of red cells in the patient sometimes results in the development of alloantibodies to red cells. This may cause difficulty in finding phenotypically compatible red blood cells. It has also been observed that autoantibodies to red cells may occur in this setting, which appear to be directed against an antigen, called N, within a red blood cell blood group

18

Table 18.1. Blood transfusion consideration in renal disease

I.   Potential transplant candidates

Leukoreduced, preferably prestorage leukoreduced RBC to prevent HLA alloimmunization and CMV transmission by transfusion.

II. Chronic Dialysis

    (a)   Multiple alloantibodies to red blood cells may develop

    (b)   Autoantibodies to red blood cells can occur

    (c)   Controversy regarding washed red blood cells

III. Prophylaxis or management of bleeding disorders

    (a)   DDAVP

    (b)   Cryoprecipitate

    (c)   Conjugated estrogens

    (d)   RBC transfusion to Hct > 35

    (e)   Recombinant human erythropoietin

IV. Management of hemolytic uremic syndrome (HUS)

    (a)   Therapeutic plasma exchange with either frozen plasma or cryosupernatant plasma as exchange fluid

    (b)   Staphylococcal protein A immunoabsorption for drug induced HUS

---

system, called the MNSS system. This antigen (Nf) may be modified by the formaldehyde used to sterilize the dialysis tubing. It is thought that this autoantibody (Nf) does not cause hemolysis in vivo, but it complicates the normal procedures for providing compatible red cells. A third consideration in patients on chronic dialysis is the occasional request for washed red blood cells. In general, the usefulness of washed red cells in this setting is related to the removal of the supernatant potassium. The supernatant concentration of K^+ is about 40 mEq/L at the end of storage, but the absolute amount given per unit of red cells rarely exceeds 5 mEq. It is best whenever possible, therefore, to transfuse patients on dialysis, which will allow for the removal of any transfused potassium. If the patient is between dialysis treatments, however, and receiving large numbers of red blood cells over a short time period (e.g., acute hemorrhage), the potassium load delivered could be excessive. This is only likely in the context of massive transfusion of such patients. Irradiated blood (Chapter 37) is best avoided, if at all possible, particularly in this

18

setting. It is preferable to transfuse fresh red cells less than 10 days old, in which the K^+ concentration rarely exceeds 10 mEq/L, if multiple red cell transfusions are given over a short time period.

Patients with renal disease also impact the blood bank on account of the bleeding disorder associated with uremia. Patients with uremia have an acquired bleeding tendency, the underlying cause of which is not clear, but thought to be related to platelet dysfunction. A clinical surrogate marker for the severity of this bleeding disorder has been the bleeding time. Strategies, therefore, which are known to shorten the bleeding time have been clinically justified in ameliorating the bleeding disorder of uremia. These strategies are outlined in Table 18.1, in more detail in Table 18.2, and are dealt with more extensively in Chapter 23. Desmopressin (DDAVP) will shorten the bleeding time within one hour of intravenous administration, but the duration of action is only 6-8 hours. A subsequent dose may be necessary, if bleeding continues. It is most often used prophylactically prior to surgery or an invasive diagnostic procedure. Although multiple dosing of DDAVP is known to result in the phenomenon of tachyphylaxis (reduction in effect with repeated doses) in Hemophilia A or von Willebrand's disease, it is unknown if this occurs in renal failure. Subcutaneous or intranasal DDAVP (the latter in higher dose) may also be administered, though there is less data available on the efficacy of DDAVP administered using these routes. Cryoprecipitate has also been shown to decrease the bleeding time. It is important to appreciate that cryoprecipitate has a delayed onset of peak action, as judged by the bleeding time, i.e., approximately 6-18 hours after administration. The effect, however, may persist for 1-2 days. The time relationship, therefore, between administration and benefit differs between DDAVP and cryoprecipitate. The dose of cryoprecipitate traditionally has been 10 units. Repeated doses could theoretically be administered, although there is no data available with regard to efficacy of multiple doses. Conjugated estrogens given by mouth (premarin 5 mg po) have also shown benefit in uremia. The onset of action is much later than either cryoprecipitate or DDAVP, occurring 24-48 hours after oral administration. The duration of action, however, may persist for 2-4 days and this strategy is useful if there is a planned elective procedure. In uremic patients with a prolonged bleeding time, there is an inverse relationship between the prolonged bleeding time and the hematocrit, and therefore, simply transfusing red blood cells to increase the hematocrit has been shown to normalize the bleeding time. This will usually occur when the hematocrit exceeds 35. If the bleeding time is a true surrogate marker for clinical bleeding events, then the transfusion of red cells should be effective in preventing excessive bleeding. Related to this is the use of erythropoietin to increase red cell mass in uremia; the resulting increased hematocrit will decrease the bleeding time. The use of allogeneic platelets in the management of the bleeding disorder of uremia is not known to be beneficial (i.e., shortening of the bleeding time has not been shown). This practice has probably arisen since the uremic defect is a "platelet defect". However, the "platelet defect" is an extrinsic platelet defect, resulting from the uremic environment, and allogeneic platelets transfused to the uremic patient will be exposed

18

Table 18.2. Treatment of the bleeding disorder in uremia

Agent	Dose	Onset of Action (hours)	Peak Action (hours)	Duration of Action (hours)
DDAVP	0.3 – 0.4 µg/Kg in 50 ml saline over 30 minutes	1/2	1-2	6-8
Cryoprecipitate	10 units IV	1	6-8	24-48
Conjugated Estrogens	Premarin 5 mg p.o. QID	24	48-72	?
Red Cell Transfusion	To achieve Hct > 35	–	–	–
rh Epo	To achieve Hct > 35	–	–	–

to the same environment. Therefore, the use of allogeneic platelets is not reasonable, has no empiric clinical justification and is wasteful.

The hemolytic uremic syndrome (HUS) is an acute renal failure in which fragmented red cells are a prominent feature. This entity occurs predominantly in children. Although some cases of HUS resolve spontaneously, plasma exchange in a dose of 1-2 plasma volume exchange on a daily basis until clinical improvement is employed in severe cases. The exchange fluid in HUS has conventionally been fresh frozen plasma, and the role of cryosupernatant as an exchange fluid (which is gaining popularity in thrombotic thrombocytopenic purpura) remains unclear (Chapter 40). In the adult population, HUS can occur in association with drugs such as mitomycin C in the treatment of cancer or cyclosporine A in the posttransplant population. In this setting, HUS is sometimes treated with staphylococcal protein A immunoabsorption therapy in which both IgG and IgG-complexes are selectively removed from plasma (Chapter 40).

18

Blood Transfusion in Medicine V: Acute Gastrointestinal Bleeding

Gastrointestinal bleeding as a clinical entity accounts for a significant proportion of all red blood cell transfusions and such patients require a prompt response in component availability from the Blood bank. Although acute spontaneous gastrointestinal bleeding has some similarities to transfusion problems seen in patients with massive trauma (Chapter 14), patients with trauma undergoing massive transfusion are usually hemostatically normal prior to the onset of the trauma, and hence, management with crystalloids and red cells may suffice until a large loss of intravascular volume has occurred. In contrast, patients with acute gastrointestinal bleeding often have an associated underlying coagulopathy at the time of presentation, and this may require early treatment with plasma or platelets in addition to any red cells transfused. Furthermore, some of these patients may have been previously transfused, and alloantibodies to red cells may be present. This may cause an unacceptable delay in making phenotypically matched red cells available resulting in the bypassing of normal procedures with associated increased risk.

Important considerations regarding the blood transfusion support of patients with acute gastrointestinal bleeding are shown in Table 19.1. First, the requirements for a large volume transfusion, arbitrarily in excess of 10 units per 24 hours, identify a high risk population with a high mortality rate. Surgery for these patients should be anticipated, making further demands for red blood cells and other components.

Many patients presenting with acute gastrointestinal bleeding have an underlying coagulopathy. Although the underlying coagulopathy may not in itself have precipitated the bleeding, as a definable anatomical abnormality may be present, the coagulopathy may exacerbate the bleeding. In addition, moderate red cell transfusion (4-6 units) may exacerbate the coagulopathy by dilution of clotting factors. Examples are patients with liver disease who develop upper gastrointestinal bleeding from esophageal varices; patients anticoagulated with warfarin who present with lower gastrointestinal bleeding; patients taking aspirin presenting with upper gastrointestinal bleeding; or patients with hypersplenism and associated thrombocytopenia. It is important to assess the presence and severity of a coagulopathy in these patients by measurement of the prothrombin time (PT) and the platelet count. The PT may be only minimally prolonged at the time of presentation. However, as these patients are transfused with relatively small volumes of red cells, (e.g., 4-6 units), a significant dilution of clotting factors will occur, unlike hemostatically competent persons. In these patients, early resuscitation with fresh

Clinical Transfusion Medicine, by Joseph D. Sweeney and Yvonne Rizk. © 1999 Landes Bioscience

Table 19.1. *Blood transfusion considerations regarding acute gastrointestinal bleeding*

I. Large volume use (10 units/24 hours) may indicate the urgent need for surgery.

II. Complicating underlying coagulopathies:

    a) Clotting factor deficiencies due to liver disease or use of warfarin.

    b) Platelet disorders such as thrombocytopenia from liver disease or hypersplenism or platelet dysfunction from aspirin.

    c) Early development of dilutional coagulopathy with modest (4-6 units) red cell transfusion.

frozen plasma at a dose of 10-15 ml/Kg is recommended. Platelet transfusion (1 unit/10 Kg) presents a more complicated decision, but would be good practice if significant thrombocytopenia ($< 50 \times 10^9$/L) is present. The role of platelet transfusion in acute gastrointestinal bleeding in patients taking aspirin is less clear. For these patients, a lower dose (3-4 units) could be given regardless of body weight, as there is evidence that only a small subpopulation of normal platelets will reverse the aspirin effect.

Patients with thrombocytopenia due to hypersplenism and who are actively bleeding require special consideration. Platelet transfusions in standard dose are unlikely to be beneficial for these patients. The most useful approach is surgical intervention to manage the anatomical site of bleeding or other kinds of manipulations, such as insertion of a tube. In the extreme situation, however, platelet transfusion should be given at a higher dose of (2-3 units/10 Kg) particularly if the platelet count is very low ($< 40 \times 10^9$/L) and there is a likelihood of further dilutional coagulopathy from surgical bleeding.

Blood Transfusion in Medicine VI: Patients Infected with Human Immunodeficiency Virus

The potential for human immunodeficiency virus (HIV) to be transmitted by blood transfusion has had a huge influence on blood transfusion practices since the 1980s. Unrelated to this, HIV infected individuals in the later stages of the disease can develop a significant red cell transfusion requirement.

When azathymidine (AZT) was initially used in HIV infected patients, the doses given were higher than those currently prescribed and anemia, requiring transfusion, was a frequent occurrence. Lower dose AZT is less commonly associated with anemia. By the late 1980s, anemia requiring transfusion in most instances was due to anemia of chronic disease related to mycobacterial avium infection (MAI) in the bone marrow. However, by 1990, recombinant erythropoietin (rh EPO) was approved for use in the anemia of HIV and resulted in a considerable reduction in the number of red cell transfusions. The recent availability of protease inhibitors, and combination therapy, which includes protease inhibitors, appears to have had a very significant effect on reducing red cell transfusion in these patients. Predictions that the HIV epidemic would result in a large increase in red cell transfusion requirements may not be borne out, if improvements in hemoglobin due to erythropoietin and protease inhibitors are sustained.

Most patients with HIV-1 infection are CMV (cytomegalovirus) seropositive and, hence, CMV transmission by blood transfusion is only rarely a consideration. It is important however to avoid primary CMV transmission by blood transfusion in the rare CMV negative patient and CMV low risk products (Chapter 38) should be given to a patient until the CMV status is known. This is most conveniently achieved using red blood cells leukoreduced by filtration.

The use of leukoreduced blood products in CMV seropositive HIV infected patients is an area of current investigational interest. A multicenter study is being performed (Viral Activation Transfusion Study) which is designed to evaluate whether patients with HIV who are transfused with red blood cell products show increased activation of their HIV disease caused by transfused allogeneic leukocytes in the red cell product. Patients in this study are randomized to receive either leukoreduced or nonleukoreduced blood products. Information regarding the results of this study will not likely be available until late 1999. In the interim, it is the practice of some physicians to use leukoreduced blood in HIV patients, regardless of CMV status, but at this time, this approach must be considered arbitrary.

There has been variation in practice with regard to use of irradiated blood products (Chapter 37) in patients with HIV. Transfusion associated graft versus

---

Table 20.1. Blood transfusion considerations in HIV infected patients

1. Anemia is a late event; role of recombinant erythropoietin and protease inhibitors.

2. Use of CMV risk reduced blood products for CMV negative or CMV status unknown recipients.

3. Use of leukoreduced blood products investigationally to prevent activation of infected T lymphocytes.

4. Use of irradiated blood products is not indicated.

5. Thrombocytopenia and the need for allogeneic platelet transfusions.

---

host disease (TA-GVHD) occurs rarely, if at all, in HIV infected individuals. There is little justification, therefore, for blood products to be routinely irradiated for this patient population. This observation is interesting since patients with HIV-1 infection have profound depression of cell-mediated immune deficiency and would, theoretically, be at risk for TA-GVHD. One speculation is that the viable transfused allogeneic leukocytes (CD4+ cells) become infected with HIV-1. As these allogeneic leukocytes become stimulated by host antigens and attempt to multiply and cause rejection of host tissue (graft versus host), they become compromised by the HIV-1 virus and replication is attenuated. This allows the opportunity for host immunocytes to reject the allogeneic leukocytes thus preventing TA-GVHD. Despite this, some centers continue to irradiate blood products routinely for patients with HIV, but this is not justifiable based on the empiric clinical experience.

Thrombocytopenia in HIV infection is related partly to infection of the megakaryocyte by the HIV virus but is also due to a decrease in platelet survival, analogous to immune thrombocytopenic purpura (ITP). Bleeding, however, is not common and the thrombocytopenia frequently responds to either steroids or intravenous gammaglobulin. Platelet transfusions are best avoided, unless a hemostatic challenge, such as surgery or an invasive procedure, is imminent (see Chapter 28).

Blood Transfusion in Medicine VII: Hereditary and Acquired Bleeding Disorders

Separating hereditary from acquired bleeding disorders is important as the treatment approaches differ between these two groups of patients. Hereditary disorders are far less common than the acquired bleeding disorders.

HEREDITARY BLEEDING DISORDERS

A list of the common hereditary bleeding disorders is shown in Table 21.1, but it is important to appreciate that outside of the context of specialized hemophilia centers the only common hereditary disorder encountered in practice is von Willebrand's disease. Patients with von Willebrand's disease, hemophilia A and hemophilia B account for approximately 95% of all the hereditary bleeding disorders. Since hereditary disorders are uncommon, few physicians are knowledgeable about the principles of care, and it is important that early consultation be obtained from a hematologist experienced in the management of these bleeding disorders. Only general guidelines are presented here for the emergency management of acute bleeding episodes in these patients.

As shown in Table 21.2, distinction between the common hereditary bleeding disorder is essential, since treatment approaches differ and laboratory monitoring of response requires specific assays. Information is often available from the patients themselves, although objective data regarding the correct diagnosis, such as discussion with the patient's physician or access to records, is desirable prior to initiating treatment. The urgent clinical circumstances and the complicated nature of these tests usually precludes confirming the diagnosis using laboratory testing in the acute setting.

Treatment of patients with hereditary bleeding disorders is outlined in Table 21.3. Patients with the common type of von Willebrand's disease (so-called type 1) respond well to desmopressin (DDAVP) in the dose indicated. When large increases in von Willebrand factor are required, as in the more severe types of von Willebrand's disease (type 3) or in certain subtypes of vWD type 2, the use of a concentrate which contains von Willebrand factor may be appropriate. A minimum increase of 50% should be sought, with 100% for life threatening situations. For an adult, this will require a dose in the range of 3,000-5,000 units. Patients with hemophilia A are optimally treated with a recombinant factor VIII, although plasma derived factor VIII is still available. Acute minor bleeding episodes, such as

Table 21.1. Hereditary bleeding disorders, arranged in order of frequency

1. von Willebrand's disease

2. Hemophilia A

3. Hemophilia B

4. Factor XI and Factor V Deficiency

5. Qualitative Platelet Disorders *requiring treatment* with platelet transfusions or other agents (Chapter 23)

Table 21.2. Laboratory differentiation of the common hereditary bleeding disorder

	FVIII:C	FIX:C	vWF: Ag	vWF: RCF
von Willebrand's disease	↓	N	↓	↓
Hemophilia A	↓	N	N	N
Hemophilia B	N	↓	N	N

FVIII:C = Factor VIII Clotting Activity
FIX:C = Factor IX Clotting Activity
vWF:Ag = von Willebrand Factor Antigen
vWF:RCF = von Willebrand Factor Ristocetin Cofactor Activity
N = Normal; ↓ = Decreased

Table 21.3. Suggested treatment of hereditary bleeding disorders for bleeding or prophylaxis prior to an invasive procedure

von Willebrand's disease:

    DDAVP   0.3-0.4 µg/Kg in 50 ml saline over 20 minutes

    Humate P   0.5 U/Kg per 1% increase in von Willebrand Factor

Hemophilia A:

    For mild hemophilia (baseline factor VIII:C > 0.05 U/mL or 5%)

    DDAVP   0.3-0.4 µg/Kg in 50 ml saline over 20 minutes

    Factor VIII:C   0.5 U/Kg per 1% increase in FVIII:C

Hemophilia B:

    Factor IX:C   1 U/Kg per 1% increase in FIX:C

Platelet Disorders:

    DDAVP   0.3-0.4 µg/Kg in 50 ml saline over 20 minutes

    Cryoprecipitate   10 Bags (units) IV

    Allogeneic platelets 1 U/10 Kg

acute joint bleed (hemarthrosis), should be treated with a minimum of 15% increase. A 100% increase is urgently required for the most severe bleeding episodes, such as intracranial, spinal cord or gastrointestinal bleeding. Patients with hemophilia B (factor IX deficiency) until recently were treated with a plasma derived factor IX. A recombinant factor IX has now become available. The increases in factor IX are the same as for factor VIII:C. It should be noted that the suggested dose for factor IX is twice that of factor VIII. This is because the low molecular weight factor IX protein (54 kD) is distributed in a larger volume than the higher molecular weight factor VIII molecule (385 kD).

The most important step in management, however, is consultation with a hematologist experienced in the care of the hereditary bleeding disorders because of the complexities in decision making regarding product type, dosing and scheduling of treatment.

Factors V or XI deficiency and the hereditary platelet disorders requiring treatment are less commonly encountered in practice, partly because they tend to produce mild bleeding manifestations. Factor V and factor IX are managed with fresh frozen plasma, 10-15 ml/Kg. Platelet transfusions should be used sparingly in the hereditary platelet disorders since other approaches may be effective (Table 21.3) and alloimmunization to platelet antigens may occur. For bleeding in some situations, such as Glanzmann's disease, platelets may be the only approach.

ACQUIRED BLEEDING DISORDERS

Differences between hereditary and acquired bleeding disorders are shown in Table 21.4. The most important point is that the acquired bleeding disorders are more common and are frequently seen in hospitalized patients, particularly in the postoperative or intensive care unit setting. The underlying coagulopathy is often less well defined and treatment is empiric. There is less predictability with regard to the clinical response to treatment. Therefore, in general, the treatment of acquired bleeding disorders is much less satisfying and, often, very controversial.

The common acquired bleeding disorders are shown in Table 21.5. It is evident that they are a heterogeneous group of disorders. The most common of these disorders, however, are liver disease, vitamin k deficiency or antagonism (such as oral anticoagulants), disseminated intravascular coagulation (DIC) and dilutional coagulopathy. Laboratory distinction between these entities can be difficult, but specific clotting factor assays can help facilitate the distinction as shown in Table 21.6. One of the more difficult distinctions is between patients with liver disease and dilutional coagulopathies. Measurements of factor VIII:C is sometimes helpful in this setting, as patients with acute dilutional coagulopathy have low levels of factor VIII:C whereas, patients with liver disease tend to have normal or high levels of factor VIII:C. Distinguishing DIC from liver disease is sometimes impossible, since DIC may occur in association with liver disease. The easiest coagulopathy to firmly diagnose is that of vitamin k deficiency. This is important

Table 21.4. Differences between the common hereditary and acquired bleeding disorders

	Hereditary	Acquired
aPTT	Prolonged	Prolonged or normal
PT	Normal	Prolonged
Coagulation Defect	Single factor; Well defined	Multiple factors; Poorly defined
Family History	Often present	Absent
Therapy	Guided by Specific assays	Usually empiric
Clinical Response to Therapy	Good; tends to be predictable	Uncertain; often unpredictable
Clinical Setting at Time of Consultation	Frequently out-patient; stable patient	Usually in-patient; unstable patient urgent/emergency care

aPTT = activated partial thromboplastin time;
PT = prothrombin time

Table 21.5. Acquired bleeding disorders

1. Liver disease

2. Vitamin k Deficiency or Use of Oral Anticoagulants

3. Disseminated Intravascular Coagulation (DIC)

4. Dilutional Coagulopathy

5. Medication Associated Platelet Defects

6. Acquired Thrombocytopenias

7. Uremia

8. Extracorporeal Coagulopathy

9. Primary Fibrinolysis

10. Heparin Associated

11. Acquired Hemophilia A or Acquired von Willebrand's disease

Table 21.6. Laboratory differentiation of the common acquired bleeding disorders

Fibrinogen	FV:C	FX:C	FVIII:C	
Liver disease	↓	↓	↓	↑
Vitamin k Deficiency	↑ or N	N	↓	↑ or N
Disseminated Intravascular Coagulation	↓	↓	N	↓ or ↑
Dilutional Coagulopathy	↓	↓	↓	↓

FV:C = Factor V Clotting Activity
FX:C = Factor X Clotting Activity
FVIII:C = Factor VIII Clotting Activity
N = Normal
↑ = Increased
↓ = Decreased

since vitamin k deficiency has a specific treatment and does not necessarily require blood transfusion.

Treatment of the acquired bleeding disorders is shown in Table 21.7. Blood transfusion is only important in some situations. Plasma in a dose of 10-15 ml/Kg is appropriate treatment in patients with liver disease or warfarin overdose who have active bleeding or if an urgent invasive procedure is imminent. A controversy surrounds the use of plasma in patients with liver disease in order to correct minimally prolonged prothrombin times. It is now appreciated (see Chapter 29) that the use of plasma in patients with mild hemostatic defects is not useful; only patients with more pronounced defects are at increased risk of bleeding.

The treatment of *vitamin k deficiency* is with low doses of vitamin k, as little as 2 mg given slowly intravenously or subcutaneously. Return of the PT to normal should be anticipated in approximately 4-6 hours although clotting factors will take longer to return to normal levels. In patients taking *warfarin*, vitamin K may be given orally if the patient is not actively bleeding or intravenously if active bleeding is present or an imminent invasive procedure is anticipated. The dose of vitamin k is larger than that required in vitamin k deficiency, and the intravenous route is, therefore, commonly used. Idiosyncratic anaphylactic reactions were reported the in past to intravenous vitamin k when given as an i.v. bolus. Vitamin k should be infused slowly, therefore, over approximately 30 minutes. Return of the prothrombin time to normal after parenteral vitamin k is approximately 5-6 hours in patients who are vitamin deficient, but with warfarin overdose this is sometimes less predictable and needs monitoring. Multiple doses may occasionally be required.

Active bleeding in the acquired platelet defects, such as the acquired thrombocytopenias, or bleeding in association with extracorporeal bypass are

Table 21.7. Treatment of acquired bleeding disorders

Liver disease:	*Plasma, 10-15 ml/Kg
Vitamin k Deficiency:	Vitamin k: 2 mg slowly IV or SC
Warfarin Overdose:	Vitamin k 10-50 mg slowly (30 minutes) IV *Plasma, 10-15 ml/kg if active bleeding
Warfarin: Reversal Prior to Elective Procedure	Oral Vitamin K to reduce INR to 1.6 Dose: 5-10 mg po, depending on initial INR
Dilutional Coagulopathy:	*Plasma, 10-15 ml/kg
Platelet Defects:	Platelets, 1 U/10 Kg IV - maximum dose 6 units (unless hypersplenism or ITP)
Extracorporeal Coagulopathy:	Platelets, 1 U/10 Kg IV - Maximum dose 4-6 units
Primary Fibrinogenolysis:	Cryoprecipitate 10 units Aminocaproic acid 4 g po q4-6 h or 1 g/h IV Tranexamic Acid 1 g po q6 h
Heparin Associated:	Protamine Sulphate (1 mg = 100 u heparin)

* To convert to "units", divide the volume of plasma required by 200

managed by transfusion of allogeneic platelets. Primary fibrinogenolysis is best heated with cryoprecipitate to replace fibrinogen (< 100 mg/dl) and concurrent antifibrinolytic therapy, although antifibrinolytic therapy may suffice as sole treatment. The diagnosis of primary fibrinogenolysis should be firm, however, and DIC (secondary fibrinolysis), should be excluded. The diagnosis of primary fibrinogenolysis is indicated by low fibrinogen, elevated fibrinogen degradation products (FDPs) and normal cross-linked fibrin degradation products (XDPs). Heparin associated bleeding is occasionally encountered with unfractionated heparin but with the increasing use of low molecular weight heparin, this is less common. Heparin excess usually shows a disproportionate prolongation of the aPTT relative to the prothrombin time and treatment is the administration of intravenous protamine sulfate. Blood products should not be administered.

Blood Transfusion in Medicine VIII: Autoantibodies to Red Cells and Platelets

The transfusion management of patients with autoantibodies to red cells or platelets complicates normal compatibility testing for these patients.

RED CELL AUTOANTIBODIES

A classification of red cell autoantibodies is shown in Figure 22.1. Red cell autoantibodies are arbitrarily divided into "cold" and "warm" antibodies, but the distinction is not absolute. Cold antibodies are antibodies which preferentially agglutinate red cells at low temperatures. They characteristically agglutinate red cells at 4°C and at room temperature (22°C), but tend not to cause agglutination at 37°C. Warm antibodies on the other hand tend to be inactive at room temperature but do cause agglutination at 37°C. Cold antibodies are mostly IgM antibodies and, therefore, may cause intravascular hemolysis due to complement fixation. Hemoglobinemia and hemoglobinuria are common. Warm autoantibodies are almost all IgG antibodies. Warm antibodies tend to cause predominantly extravascular hemolysis. Hemoglobinemia or hemoglobinuria is rare. Regardless of the type of hemolysis, either condition may result in severe anemia and give rise to difficulties with compatibility testing and, hence, delay in the availability of phenotypically compatible red blood cells.

The major considerations with regard to transfusing red cells in patients with red cell autoantibodies are shown in Table 22.1. The first consideration is distinction between the presence of an autoantibody or alloantibody(ies). The test, which detects antibody or complement bound to the surface of the red cells, is called the direct antiglobulin test, or more commonly, the direct coombs test. This test should be positive in the absence of recent (< 3 months) transfusion. If the antibody is present in the plasma, it should lack antigen specificity and should agglutinate all cells (called a panagglutinin). Antibody bound to the red cell membrane can be displaced (eluted) using chemicals or strong acids. This cell bound antibody should also show the same characteristics of the plasma antibody (i.e., a panagglutinin). The second consideration after establishing the presence of an autoantibody is the detection of an additional possible underlying alloantibody(ies). In practice, much of the blood bank's work focuses on this second question, and in this regard a history of previous transfusion or pregnancy is important since these patients are potentially at risk for the presence of underlying alloantibodies.

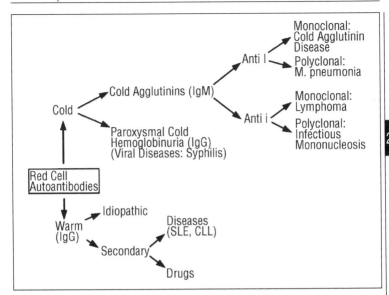

Fig. 22.1. Classification scheme for red cell autoantibodies

Table 22.1. Considerations regarding red cell transfusion in patients with red cell autoantibodies

1. Is there a prior history of blood transfusion or pregnancy?

2. How low is the hemoglobin/Hct? Is the transfusion critical?

3. Is the antibody on the red cell membrane (direct coombs), in the serum/plasma (indirect coombs), or both?

4. Transfuse leukoreduced red cells (Chapter 36) to avoid nonhemolytic reactions.

5. Transfuse slowly, if possible, with vigilance for clinical symptoms of hemolysis.

6. Use blood warmers if available, for cold antibodies.

In searching for red cell alloantibodies in patients with red cell autoantibodies, the blood bank frequently engages in a number of sophisticated techniques, most of which are time consuming. These techniques involve attempts to absorb the autoantibody from the patient's plasma in order to detect and/identify the presence of an alloantibody. Often, this is unrewarding. Considerable delay of many hours can result before the testing procedures are completed. Patients with warm antibodies can generally be front typed for the ABO system and for most of the antigens within the Rhesus system (Chapter 6). Red cells can be made available which are phenotypically compatible with the major antigens within these systems. If the autoantibody is only detected on the red cell membrane (positive direct coombs test), and absent in the serum (negative indirect coombs test), then the procedures are less time consuming. If the antibody is both cell-bound and present in the serum (which is a common situation), the above considerations will apply. Red blood cells will frequently be incompatible using the standard tests. Physicians may need to sign a release form acknowledging this incompatibility. This serves an important purpose in that it reinforces the need for extra vigilance. In practice, however, most of these transfusions are well tolerated and produce the expected increase in hematocrit.

The clinical decision to transfuse these patients should be made cautiously because of the potential higher risk for the occurrence of hemolytic reactions. Patients with red cell autoantibodies should have blood transfused using a leukoreduction filter. This is to avoid potential confusion occurring during the transfusion of these patients due to transfusion reactions caused by allogeneic leukocytes (Chapter 32). The transfusion should be performed with vigilance and care, with particular careful observation for clinical symptoms of hemolysis (Chapter 32).

Patients with cold antibodies present some different considerations. The most important considerations are the *thermal range* and the *antibody titer*. Cold antibodies which react at room temperature (22°C) only (and not at 37°C) are rarely clinically significant. In addition, low titer antibodies (< 1:64) do not cause hemolysis. As cold antibodies do not react (cause agglutination) at 37°C, screening for minor blood group alloantibodies is possible and finding compatible red cells less difficult. Patients with high titer cold agglutinins may show discordancy between the ABO front and reverse type since they reverse type as group O (Chapter 6). If doubt exists regarding ABO type, transfusion with blood group O cells is most appropriate. When a transfusion is required, the patient should receive leukoreduced blood and be preferably transfused using a blood warmer.

PLATELET AUTOANTIBODIES

Platelet autoantibodies are most commonly seen in idiopathic thrombocytopenic purpura (ITP). It is important to appreciate that, although such patients may have low platelet counts (< 10 x 10^9/L), the platelets are larger in size and the hematocrit is usually normal. This differentiates these patients from other patients with thrombocytopenia, such as acute leukemia, where the platelets are normal or reduced in size and the hematocrit usually decreased. Patients with ITP may show evidence of mucosal bleeding, such as easy bruising, and sometimes epistaxis, but severe bleeding is not frequently observed and it is likely that the larger platelets and higher hematocrit are protective to the patient in this regard. Thus the threshold for the platelet transfusion in a patient with ITP is not the same as in diseases such as acute leukemia. In addition, the natural history of

---

Table 22.2. Considerations regarding platelet transfusion in patients with platelet autoantibodies

---

1. Patients often have large platelets and normal hematocrits, which may protect against bleeding.

2. How low is the platelet count and is there clinically significant active bleeding?

3. Is an invasive procedure imminent?

4. A rapid response to treatment may occur (within 48 hours).

   (a) Intravenous gammaglobulin
      2 mg/Kg in divided doses within 5 days

   (b) Anti-D (Win-Rho)
      50-75 µg/Kg as a single IV treatment

   (c) Prednisone
      1-2 mg/Kg po QD x 14 days

   Platelet transfusion should only be used for active bleeding which is severe or life threatening. The dose of platelets (number of units in the pool) should be twice to three times standard in order to achieve a predictable increase.

5. If clinically indicated, fresh, pooled platelets may be the optimal platelet product. Fresh platelets (less than 3 days) will have a lower likelihood of a transfusion reaction and a higher likelihood of achieving a platelet increase (transient) on account of the antigen heterogeneity of the pool.

22

these autoantibodies in children is spontaneous resolution and in adults there is usually a rapid response to either intravenous gammaglobulin, corticosteroids or Anti-D (Win-Rho). Avoidance of platelet transfusion is preferred, if at all possible.

If a platelet transfusion is judged appropriate, however, because of the presence of more serious bleeding or if an invasive diagnostic or therapeutic procedure is required, these patients are best transfused with a pool of fresh, random donor platelets. The dose to be transfused is largely empirical but should ordinarily be at least twice the normal dose, i.e., approximately 10-16 units of random donor platelets. The platelets are best transfused fresh since they are less likely to cause transfusion reactions and the pool of random donors is preferable to a single donor product because of antigen heterogeneity and the higher likelihood of response. Patients with autoantibodies to platelets may respond to platelet transfusion with increases in the platelet count, but the response tends to be blunted and transient (less than 3 hours). Therefore, an invasive procedure, if anticipated, should be performed within 30-60 minutes after completion of the platelet transfusion.

A unique clinical situation is the management of these patients undergoing splenectomy. Despite the fact that the platelet count is low at the initiation of surgery, it is generally advised that platelet transfusions be withheld until the splenic artery is clamped. At this point, a standard dose of platelets may be administered with reasonable expectation of an increment in the platelet count. This should allow the surgeon to complete the splenectomy without excessive hemorrhage.

22

Blood Transfusion in Medicine IX: Using Drugs to Reduce Blood Transfusion

Understanding the use of prohemostatic pharmacological agents is important, since they may be effective in reducing allogeneic blood exposure in certain patient populations. The range and types of products used varies, and the evidence for therapeutic efficacy is based on empiric clinical experience showing a reduction in bleeding in some instances and, in others, using surrogate markers for bleeding, such as the bleeding time. A classification is shown in Table 23.1.

HORMONES OR HORMONE DERIVATIVES

The most important agent in this group is desmopressin, or 8 desamino-8-D-arginine vasopressin, often abbreviated DDAVP. DDAVP was initially used in the early 1970s in the treatment of patients with mild hemophilia A and von Willebrand's disease and consistently caused a transient increase in factor VIII and von Willebrand factor. Subsequently, DDAVP was shown to shorten the bleeding time in patients with uremia and in patients with platelet storage pool disease. It was also shown to reduce blood transfusion in patients undergoing spinal fusion surgery, a procedure associated with significant red blood transfusion. In the mid-1980s, one study reported that DDAVP was effective in reducing blood transfusion in patients undergoing cardiac surgery, but subsequent clinical trials have not confirmed this observation and DDAVP is now considered to be of unproven value in reducing blood loss in cardiac surgery.

The most common use of DDAVP outside of factor VIII deficiency states (see Chapter 21) is in the treatment of acute uremic bleeding or as prophylaxis in a patient with uremia prior to an invasive procedure.

The onset of action of DDAVP is approximately 20-30 minutes after the infusion, but the peak of factor VIII is at 30-60 minutes; for uremia, the peak action (i.e., reduction in bleeding time) is at 4-6 hours. Multiple doses can be given in the factor VIII deficiency state, but tachyphylaxis (diminished response after repeated doses) may sometimes occur. Repeated doses in uremia, surgery or hereditary platelet disorders is of unknown benefit.

The second agent in this category is conjugated estrogens. Conjugated estrogens are a mixture of two different hormones and in early experiments were shown to be useful in the treatment of uremic bleeding. Intravenous premarin given daily

---

Table 23.1. Pharmacologic agents used to reduce bleeding

I. Hormones or hormone derivatives:
    A. desmopressin (DDAVP) 0.3 µg/Kg in 50 mls saline over 20 minutes
    B. conjugated estrogens
       Premarin   0.6 mg/Kg IV QD x 1-5 days
       Premarin   5 mg po q6h QD x 5 days

II. Antiproteases:
    A. Aminocaproic acid   4 g q 4-6h po
       (Amicar)   4 g IV q4h
                             or I g q6h po
    B. Tranexamic Acid   1 g po q6h
       (Cyclokapron)   0.5 g q8h IV
    C. Aprotinin   (Trasylol®):

1. Cardiac Surgery:	Full dose	Half dose
	2 MU pre pump	1 MU pre pump
	2 MU pump	1 MU pump
	0.5 MU/h	0.25 MU/h
	post pump	post pump

2. Orthotrophic Liver	2 MU bolus postinduction
Transplantation	1.5 MU/h during procedure

III. Cytokines
     A. rh Epo:
        30-50 IU/Kg TIW (dialysis); maintenance 25 IU/Kg BIW
        Target Hct 30-35
     B. TPO: Not yet licensed
     C. Interleukin 11
IV. Topical Hemostatic Agents
     Fibrin glue (Tisseel)
     Topical Thrombin
     Collagen
        1. microcrystalline (Avitene)
        2. positively charged modified (Superstat)
     Oxidized cellulose (Surgicel)

---

for several consecutive days shortened the bleeding time and showed clinical evidence of reduced bleeding in uremia. This effect had an onset several days after the infusion, but a duration of 10-14 days. More recently, premarin has been given by mouth for several consecutive days, similarly reducing the bleeding time with a concomitant reduction in clinical bleeding.

The second category are the antiproteases. These are best divided into two subgroups: The low molecular weight drugs, such as aminocaproic acid (Amicar) and tranexamic acid (Cyclokapron), have identical mechanisms of action. These drugs act primarily by inhibiting the enzyme plasmin. Aminocaproic acid and tranexamic acid differ in dosing, however. These agents have their main use in reducing mucosal bleeding, particularly oral bleeding. They have been used in other situations, such as epistaxis and in bleeding from the urinary tract. Tranexamic acid and aminocaporic has been used empirically in thrombocytopenic patients, but there is no data that either agent reduces platelet transfusion episodes or has a clinical effect in reducing bleeding. These agents have also been used immediately prior to cardiopulmonary bypass and in some studies have been shown to decrease both chest tube drainage and total blood transfusions. The other antiprotease is an agent called aprotinin. Aprotinin (Trasylol) is a 65 kD protein which is extracted from bovine lung. Aprotinin inhibits a number of enzymes, particularly plasmin, kallikrein, and activated protein C. Aprotinin has found clinical application in several situations. In multiple studies both in Europe and the United States, aprotinin has been shown to reduce bleeding and allogeneic transfusion in cardiac surgery. The initial dose of aprotinin (full dose or Hammersmith dose) used approximately 6 million units (Table 23.1). A half-dose regimen has been shown to be equally efficacious in reducing blood transfusion. A related interesting observation is a reduction in postoperative stroke in patients undergoing cardiac surgery treated with aprotinin. This is achieved, however, only by the use of the full dose regimen. Aprotinin has also been used in orthotopic liver transplantation, where a substantial decrease in total blood transfusion has been reported. There is emerging data for the use of aprotinin to reduce blood loss in orthopedic surgery. There are isolated reports of the use of aprotinin during acute bleeding episodes in patients with thrombocytopenia refractory to platelet transfusions, but neither the indication nor the dosage is well established.

The third group of drugs is *cytokines*. Of these, recombinant human erythropoietin (rhEpo) is the most important. rhEpo is primarily used in patients on chronic dialysis in order to increase hematocrit and reduce symptoms of anemia, but the increases in hematocrit are associated with a shortening of the bleeding time. A variety of cytokines influence platelet production and may be useful in thrombocytopenia. These are granulocyte-monocyte-colony stimulating factor (GM-CSF), Interleukin-3 (IL-3), IL-11 and thrombopoietin (TPO). TPO is a recently cloned cytokine which may prove useful in the treatment of patients with thrombocytopenia due to chemotherapy or bone marrow transplantation, but early results from clinical studies are disappointing. This agent has not, as yet, been approved for this clinical indication. Interleukin-11 has been approved, however, for this indication in the United States.

A fourth group of agents are the topical hemostatic agents. The most important agent in the use of this group is fibrin glue. Fibrin glue is a generic name which refers to a variety of preparations which are essentially concentrates of fibrinogen and/or fibronectin. The product may be either autologous or allogeneic, although usually, it is allogeneic. A lyophilized product has recently been approved for use in the United States (Tisseel). Fibrin glue can be a valuable topical agent in the treatment of superficial surface bleeding, such as in redo cardiac surgery. It is also valuable in trauma with liver laceration, where it has been shown to be effective in reducing bleeding and in neurosurgery or vascular surgery. Topical thrombin is another agent which has been used for minor superficial and often mucosal type bleeding. Collagen preparations have also been applied topically to control bleeding in surgery and two types of preparations are available: A microcrystalline powdered form and a positively charged modified collagen form. A proven role for either of these agents in reducing bleeding, and hence in potentially reducing transfusion has not been shown and neither agent is known to be superior to fibrin glue. Microcrystalline collagen has also been associated with extensive scaring. Caution needs to be exercised if these agents are applied whenever intraoperative salvage is being used, and aspiration from the site should be discontinued. Last, oxidized cellulose is a product derived by exposing cellulose to nitric oxide. This product appears to control hemostasis by trapping blood elements in a mesh. It is questionable whether this product is any more beneficial than the simple application of gauze with pressure.

23

Blood Transfusion in Obstetrics

The major blood transfusion considerations in obstetrics are shown in Table 24.1. As physiologic preparation for blood loss at the time of delivery, the blood volume of a gravid woman is 60% more than that of a nonpregnant woman resulting in a dilutional anemia. It should be emphasized that patients can tolerate moderate anemia (hematocrit 18-25%, hemoglobin 6-8 g/dl) if normovolemia is maintained. Blood transfusion is an uncommon event in obstetrics. Only 1% of vaginal deliveries require transfusion. However, as many as 18% of patients undergoing cesarean section may require transfusion. Overall obstetrical patients account for 2-4% of all red blood cells transfused in the U.S. (Fig. 4.1).

Early in the prenatal period, a pregnant woman should be evaluated for a family history of bleeding disorders or a history of blood transfusion. Routine laboratory tests should include the hemoglobin/hematocrit, ABO and D type, and antibody screen and screening for hemoglobinopathies in high risk populations.

Immunization to the D antigen in Rh negative mothers is the primary cause of hemolytic disease of the newborn (HDN). Prevention is critical and best performed, using anti-D (RhIg). The clinical indications are shown in Table 24.2. The approach is as follows:

1) Abortion, ectopic pregnancy or abdominal trauma. The Rh antigen is demonstrated as early as 38 days in fetal red blood cells. Treatment is a dose of 50 μg if the event occurs before 12 weeks, and 300 μg when it occurs later in pregnancy.

2) Amniocentesis. Amniocentesis performed prior to 20 weeks of gestation can produce fetomaternal bleeding of between 0.5-10 ml. The optimal treatment is the administration of 300 μg prophylactically when the father is Rh positive, without relying on the Kleihauer-Betke acid elution technique. This dose is adequate until 28 weeks gestation when a subsequent antenatal dose is administered.

3) Hydatidiform mole. The role of anti-D prophylaxis is not established; however using the above guidelines would seem prudent.

4) Late pregnancy. Pregnancy manipulation such as abdominal version and amniocentesis enhances the risk of transplacental hemorrhage. If delivery is to be accomplished within 48 hours of the amniocentesis, the administration of Rh immunoglobulin can be deferred and given only if the infant is found to be Rh D positive.

Otherwise, routine management is as follows:

1) Obtain ABO blood group and Rh (D) type and screen in the first antenatal visit.

2) For Rh (D) negative women at 28 weeks gestation, obtain an indirect Coomb's test (antibody screen); if no Rh antibodies are detected,

---

Table 24.1. Blood rransfusion considerations in obstetrics

---

1. Maternal circulation exhibits a dilutional anemia, which is a normal adaptive change. Threshold hemoglobin/Hct for transfusion may be different (Chapter 26).
2. Blood typing (ABO, D) and antibody screening should be performed early in pregnancy.
3. Prophylaxis with Anti-D, if relevant (Chapter 25) for any invasive procedure, abortion or delivery.
4. Severe acute hemorrhage generally occurs in late pregnancy or at delivery. Few patients may actually need transfusion. CMV low risk red cells are essential if transfused in early pregnancy.
5. Predeposit autologous red cells are rarely effective since predictability of bleeding (and hence, transfusion) is difficult in individual patients.
6. Thrombocytopenia may be common (5-7%) and is often mild (80-120 x 10^9/L). *Platelet transfusions are rarely, if ever, indicated.*
7. Transfusion of plasma/cryoprecipitate is rare only occurring in the context of acute massive bleeding with dilution or in obstetric-associated disseminated intravascular coagulation.

Table 24.2. Clinical indications for RhIg prophylaxis

Early Pregnancy	Late Pregnancy	Postpartum
Abortion	Fetomaternal hemorrhage	Pre- or term deliveries
Ectopic pregnancy	Amniocentesis	
Hydatidiform molar pregnancy		
Amniocentesis		
Chorionic villus sampling		

---

administer 300 µg (one dose) of Rh immunoglobulin. This should provide protection for 12 weeks.

3) At delivery: Rh type and direct Coomb's test is performed on the cord blood. If the baby is Rh positive, 300 µg of Rh immunoglobulin is administered to the mother within 72 hours postpartum. If a large transplacental hemorrhage is suspected, a Kleihauer-Betke stain is done to quantitate the total bleed and 10 µg of Rh immunoglobulin given for each ml of fetal RBCs. Anti-D preparations are available for intramuscular (Rhogam, Win-Rho) or intravenous use (Win-Rho).

The main indication for elective red cell transfusion is the inherited hemoglobinopathies, i.e., sickle cell disease, Hb C disease, Hb S/C disease, Hb S/β^0 thalassemia. Treatment of these disorders with red cell transfusion is to avoid complications such as infection, to control pregnancy-induced hypertension, and to prevent and treat vascular occlusive episodes. The role of transfusion is controversial, and it has been suggested that it should be reserved for obstetric emergencies only. If transfusion is performed, the red cells should be sickle cell negative. Ideally, hemoglobin A is maintained between 20-40%, with a hematocrit in excess of 25.

Leukocyte-reduced RBCs should be transfused, to avoid reactions and prevent CMV transmission (Chapter 36).

Acute blood loss can be a sudden event in obstetrics. The causes are shown in Table 24.3. Depending on the severity (Table 24.4), transfusion may be required. In massive transfusion, (arbitrarily after 10 more units of blood have been transfused), the entire blood volume of the pregnant woman has been replaced. In such rare cases, the patient should be followed by serial assessments of clotting times and platelet counts and replacement with plasma or platelets may be necessary (Chapter 14).

Predeposit autologous blood is sometimes collected from pregnant females. This practice is rarely of benefit to the obstetrical patient since the ability to predict the rare patient needing allogeneic transfusion is difficult (Chapter 3).

24

Table 24.3. Common causes of obstetrical hemorrhage

In Late Pregnancy	Delivery & Postpartum
Abruptio placenta	Cesarean delivery
Placenta previa	Obstetric laceration
Toxemia associated	Uterine atony
	Retained placenta
	Uterine inversion
	Placenta acreta

Table 24.4. Classification of acute blood loss in obstetrics

Grade	Approximate Volume of Acute blood loss	Approximate Percentage Loss of Blood Volume	Signs, Symptoms & Treatment
1	600-1200 ml	10-20	Minimal tachycardia. Saline replacement adequate.
2	1200-1800	20-25	Increased pulse rate, elevated respiratory rate, orthostatic blood pressure change, narrowing of pulse pressure. May require blood transfusion but can be stabilized with crystalloids.
3	1800-2400	30-40	Reduction in systolic blood pressure, significant tachycardia and tachypnea. Blood transfusion is usually needed.
4	> 2400	> 40	Profound shock with no discernable blood pressure. The patient has oliguria or anuria. Blood transfusion is mandatory.

Thrombocytopenia in pregnancy is not uncommon. The causes are shown in Table 24.5. Platelet transfusions are rarely given—the only exception being severe DIC associated with amniotic fluid embolism, fetal death, or abruptio placentae. If Rhesus positive platelets are given to a Rhesus negative female, anti-D (RhIg) should be administered (50-300 µg). The platelets should be leukoreduced, preferably prestorage (Chapter 36) to avoid reactions and prevent CMV transmission.

The only common hereditary bleeding disorder in females is von Willebrand's disease. However, because of changes in coagulation factors and von Willebrand factor in pregnancy, both pregnancy and delivery are rarely complicated by bleeding. Where concern exists, DDAVP may be used, after presentation of the shoulder or the baby has been delivered by cesarean section (Chapter 21), unless contraindicated by an uncommon subtype of von Willebrand's disease, such as type 2b.

24

Table 24.5. Causes of Thrombocytopenia

1. Incidental Thrombocytopenia:
   5-7% of all pregnancies; mild thrombocytopenia in the range of 80-120 x 10^9/L
   No known increase in maternal or fetal morbidity or mortality. Does not change obstetrical management.

2. Hypertension Associated:
   Remits with early delivery.

3. Immune Thrombocytopenic Purpura:
   Uncommon. Thrombocytopenia, may be <20 x 10^9/L.
   Best treated with IVGG, if necessary.

4. Miscellaneous (rare):
   HELLP - Hemolytic anemia with elevated liver enzymes in pregnancy.
   TTP - Thrombotic thrombocytopenia purpura.
   DIC - Disseminated intravascular coagulation.

Fetal and Neonatal Transfusion

FETAL TRANSFUSIONS

The most common indication for fetal transfusion is hemolytic disease of the newborn (HDN). The most common cause of the syndrome is maternal antibodies to the Rh system, which is comprised of five common antigens, C c D E e, and other rare antigens (Chapter 6). The D antigen is the most antigenic (provokes antibody formation) and consequently is implicated most frequently. Once maternal immunization occurs, it cannot be reversed. Therefore, the best management is maternal immunoprophylaxis by the use of Rhesus immunoglobulin (RhIg or anti-D, Chapter 24).

Assessment of the severity of HDN is possible by performing amniocentesis at 26-28 weeks with measurement of bilirubin in the amniotic fluid. In the past, serial amniocenteses were performed at 1-4 week intervals followed by intraperitoneal transfusion(s). Currently cordocentesis (sampling of cord blood) is performed, which gives a direct measure of the degree of fetal anemia. Intrauterine transfusions (IUT) of red cells may be used in the third trimester. The choice of red cell product is shown in Table 25.1. IUT may be given by the intraperitoneal (IP) route or by cordocentesis. Cordocentesis is gaining favor, because of the lower complication rate and the frequent lack of success of transfusion(s) using the IP route in very severe cases of HDN.

Alloimmune thrombocytopenia occurs in approximately 1: 2000-1:5000 pregnancies. This condition is due to antibodies to platelet antigens which cross the placenta and destroy fetal platelets. It is similar to HDN, except that the antibodies are directed against platelets, rather than red cells. Maternal serum is not routinely screened, however, for platelet alloantibodies. IUT of platelets are given with second or later pregnancies, often where the first infant was born with thrombocytopenia.

NEONATAL TRANSFUSIONS

LARGE VOLUME EXCHANGE TRANSFUSION
Large volume exchange transfusion is accomplished via partial replacement of whole blood with red blood cells reconstituted in fresh frozen plasma with the hematocrit measured every 6-24 hours. Calculations are based essentially on estimated blood volumes of 80 ml/Kg in term infants and 100 ml/Kg in premature

Clinical Transfusion Medicine, by Joseph D. Sweeney and Yvonne Rizk. © 1999 Landes Bioscience

---

Table 25.1. Intrauterine transfusions

I.   Red cell products
    A. 1.) Group O, D negative, Sickledex negative.
         2.) Washed maternal red cells, if mother is group O. More caution if mother is other than group O. Crossmatched with maternal plasma or serum.

    B. Red cells should be frozen, deglycerolized or leukoreduced by filtration to prevent CMV transmission (Chapter 38)

    C. Irradiated by gamma irradiation (Chapter 37)

II. Platelet products

    A. 1.) Compatible with known platelet alloantibody
         2.) Maternal platelets

    B. Leukoreduced

    C. Irradiated by gamma irradiation

    D. Washed

25

infants. Simple whole blood exchange will replace 65% of total blood volume. Double volume will replace 85% of blood volume. The red cells should be fresh (less than 5 days) or washed because of the potassium load. In general, the procedure should take less than 60 minutes to reduce the risk of introducing infection. Aliquots of 10-20 ml are removed at a time providing that no more than 10% of the estimated blood volume is extracorporeal at any time. The indications for exchange transfusion are shown in Table 25.2.

SMALL VOLUME NEONATAL TRANSFUSION

The most common blood product transfused is red cells, in order to treat anemia due to acute blood-loss and chronic (intrauterine) blood loss. The causes are shown in Table 25.3. The red cell dose is 10-15 ml/Kg. Although fresh red blood cells in anticoagulant (CPD-RBC, Chapter 2) are frequently requested, many neonatal units transfuse red cells in additive solutions regardless of the storage age, whenever this dose is given. Group O red cells are advisable for all neonates, and should always be used for younger, low birth weight infants.

NEONATAL THROMBOCYTOPENIA

This may result from impaired production or increased destruction of platelets. Platelet transfusion is often indicated in neonates and young infants with

Table 25.2. Indications for exchange transfusion

All components used for intrauterine transfusion or in neonates of 1.2 Kg or less must be irradiated and should have a reduced risk of CMV transmission such as seronegative donors, deglycerolized, leukocyte reduced by filtration. They must be cross match compatible with maternal serum.

1. Hemolyte Disease of the Newborn (HDN)

2. Hyperbibrulinemia: (unrelated to HDN)

Exchange transfusion is mandated at a bilirubin level of 18-20 mg% by 24-48 hours of age in a metabolically unstable, full term infant, and at a bilirubin level of 25 mg% in a metabolically stable infant.

- Liver conjugation system immaturity
- Hereditary red cell disorders (e.g. spherocytosis, elliptocytosis)
- Hemoglobin synthesis disorders such as thalassemia and sickle cell disease

3. Sepsis:

The most effective modality of treatment is whole blood exchange transfusion to detoxify endotoxins with or without antibiotics.

4. Disseminated Intravascular Coagulation (DIC)

Treatment is ideally with fresh whole blood exchange transfusion to provide clotting factors and remove fibrin degradation products.

5. Polycythemia:

Hyperviscosity in this disease is directly related to neurologic impairment in the neonate. It is important to note that the peripheral hematocrit is disproportionate to the central nervous system hematocrit and should not be used as a guideline to estimate the degree of viscosity. Decision is based on both clinical and metabolic status, lethargy, hypoglycemia, and hypocalcemia.

25

Table 25.3. Causes of neonatal anemia

1. Soft tissue rupture

2. Loss of vascular integrity leading to blood loss in body cavities

3. Twin-twin transfusion

4. Fetal-maternal transplacental bleeding

5. Obstetric related blood loss such as abruptio placenta and placental tears

6. Blood sampling

platelet counts below 50 x 10^9/L (50,000 mm^3) and who are bleeding. Neonatal alloimmune thrombocytopenia at term is managed as for fetal transfusion (Table 25.1). However, in the neonate, this transfusion may occur with the first pregnancy and technically, the platelet transfusion is much simpler. The dose is 1 unit.

Granulocytic transfusions are used rarely in septic infants. Ideally exchange transfusion with fresh whole blood is effective in treating the septic neonate. However, granulocyte support may be useful in the treatment of gram-negative and Staphylococcal infections. The dose is 1 x 10^9 neutrophils in a volume of 15-20 ml. This dose and volume is approximately present in the fresh (less than 8 hours) buffy coat from a unit of whole blood, but a fresh granulocyte apheresis product is preferred, if available. The donors are usually selected to be CMV seronegative and the granulocytes are irradiated to prevent transfusion associated graft versus host disease (Chapter 37). Granulocyte products are never leukoreduced by filtration.

Fresh frozen plasma (FFP) is the treatment of choice to replace coagulation factors in a dose of 10-15 ml/Kg unless a specific concentrate is available. FFP increases clotting factor activity by 10-20%. This is used to treat bleeding in (1) hereditary coagulation factor deficiency or (2) maternal causes such as antenatal disseminated intravascular coagulation (DIC) or vitamin K deficiency. Rarely, cryoprecipitate is transfused in cases of DIC, often in conjunction with platelet transfusion. The dose is 1-2 units.

Clinical Decisions and Response Monitoring: Triggers, Targets, Functional Reserve and Threshold of Effect

The decision to transfuse any blood component should never be made exclusively on the basis of a laboratory test. In many instances, however, laboratory tests are important in guiding the appropriate use of blood components. The clinical practice of routinely transfusing patients merely on the basis of laboratory results such as a hemoglobin or hematocrit without regard for clinical symptoms has fallen into disrepute. A combination of clinical and laboratory features is the basis of good clinical judgment regarding the need for blood transfusion.

The reason is illustrated in Figure 26.1. This shows a general relationship between clinical symptomatology and laboratory test results. The abscissa, (y-axis) shows clinical symptoms. The ordinate (x-axis) shows a laboratory test which has a normal range and then increasing in the degree of abnormality. In more concrete terms, the clinical symptom could be fatigue or malaise and the laboratory test the hematocrit. Within the normal range of the hematocrit (arbitrarily 38-52), clinical symptoms such as fatigue cannot be attributed to the hematocrit. As the hematocrit becomes abnormal, (arbitrarily between 27-38), clinical symptoms due to anemia are unlikely in most patients. This is largely because of the ability of the heart to compensate by increasing cardiac output, which will likely occur as the hematocrit drops below 30. As the hematocrit drops further (21-27), some clinical symptoms may occur. This will, of course, depend on the specific clinical situation, the degree of the anemia, the rate of blood loss, and patient features such as age, physiological status, and cardiac function. The point at which clinical symptoms become evident is called the *threshold of effect*. The threshold of effect is not to be mistaken with the *transfusion trigger*. Patients with minimal symptoms of anemia who may respond to other forms of treatment, such as iron or vitamin B12 deficiency, etc. are not transfused at the threshold of effect. In addition, bedridden patients without any expectation of an immediate increase in exercise need, not be transfused at this threshold. As the degree of abnormality worsens, however, a point is reached where the clinical symptoms justify a blood transfusion. This point for any individual patient is known as the *transfusion trigger*. When a decision is made to transfuse the patient, consideration must be made as to the expected outcome. An adequate dose of the blood product should be given (in this case, the volume of red cells to be prescribed) in order to achieve a *target*

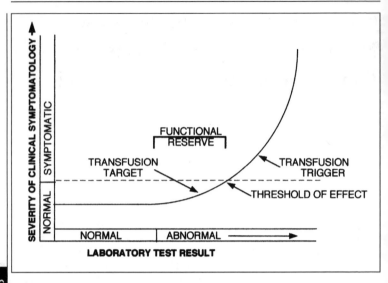

Fig. 26.1. Clinical transfusion decision-making. Theoretical relationship between the severity of clinical symptoms and the degree of abnormality of a laboratory test result.

posttransfusion result. This should be below the threshold of effect in order to allow a safety margin in any individual patient and to ensure that there will be a satisfactory outcome (improvement in symptoms) from the blood transfusion. The transfusion target, however, need not be within the normal range. The degree of abnormality which an individual patient can sustain once laboratory results begin to shift from the normal range until it reaches the threshold of effect is called the *functional reserve* for that particular patient. Functional reserve is due to a compensatory mechanism, such as increased cardiac output, increased red cell 2, 3 diphosphoglyceric acid, etc.

These concepts are of importance in making appropriate clinical decisions with regard to the transfusion of individual patients. Applying these concepts to platelet transfusions is as follows: As the platelet count drops slightly below the normal range of 140×10^9/L, clinical bleeding will not occur, and the count may decrease to 30×10^9/L or lower before an increased risk of minor spontaneous clinical hemorrhage becomes evident (threshold). However, the transfusion trigger i.e., the decision to transfuse platelets, will be much lower than the 30×10^9/L, e.g., for example, 10×10^9/L. Once a decision is made to transfuse, the dose of platelets should result in a $20\text{-}40 \times 10^9$/L increase in the platelet count, i.e., the transfusion target will be beyond threshold of effect. It can be seen, therefore, from this example that the functional reserve in platelets is very large and extends well into the abnormal range. A further change in immune thrombocytopenic purpura (ITP) is that the platelets are larger (Chapter 22). Therefore, in this condition, even very low platelet counts are tolerated for long intervals without apparent significant bleeding. These concepts can also be applied to the transfusion of fresh frozen plasma in a patient with liver disease. As the prothrombin time (PT) prolongs slightly, all available data indicates that there is little or no increase in clinical bleeding. At some arbitrary prolongation of the prothrombin time, a slight increase in bleeding risk of no clinical significance could become manifest (threshold of effect) if an invasive procedure were performed. The transfusion trigger should be beyond the threshold. Plasma at a dose of 10-15 ml/kg will likely result in a shortening of the PT below this threshold. Note that the transfusion target is not within the normal range. There is a common misconception in attempting to achieve a prothrombin time within the normal range prior to an invasive diagnostic or therapeutic procedure. In more concrete terms, using a thromboplastin with an ISI of 2.0, the upper normal PT could be 13 seconds, then the functional reserve is probably 14-16 seconds, the threshold of effect at 16 seconds, the transfusion trigger 18 at seconds and the transfusion target, 15 seconds.

If compensatory mechanisms are compromised, the above principles do not change, but the critical values may shift. This is illustrated in Figure 26.2. In this figure, the theoretical relationships between fatigue, a symptom of anemia, and hematocrit in two hemodynamically stable-iron deficient subjects aged 20 and 80 years is shown. The symptomatic threshold for the 20-year-old may be a hematocrit of 20; for the 80-year-old, at a hematocrit of 30. The transfusion trigger, however, for the 20-year-old could be a hematocrit of 10-14; for the 80-year-old, 24-27. The above assumes that there is no imminent threatening acute blood loss.

26

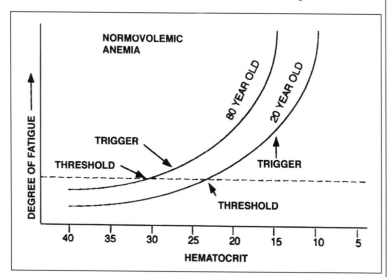

Fig. 26.2. Theoretical relationship between fatigue and degree of abnormality of the hematocrit in 80 year old and 20 year old males.

In monitoring the laboratory response to transfusion, for red blood cells, the hematocrit can be measured at 1-24 hours posttransfusion in the absence of on-going blood loss. For platelets, the increment is measured at 10-60 minutes posttransfusion to measure 'recovery'; and at 18-24 hours to estimate survival. For plasma, the prothrombin time or activated partial thromboplastin time can be measured after plasma have undergone blood volume equilibration, usually after 3 minutes. However, 10-15 minute postplasma transfusion would be reasonable. Some clotting factors such as factor VII have short a half life (3 hours) and a low molecule weight (factors II, VII, IX, X), such that they will equilibrate with the extravascular space. Therefore, the beneficial effect of plasma transfusion as measured by a shortening of the prothrombin time tends to be short lived.

26

Red Blood Cells:
Indications and Dosing

Red blood cells are manufactured from a whole blood donation by the removal of plasma. Most of the white cells (approximately 90%) and platelets remain with the red blood cell component unless a platelet concentrate is manufactured from the blood donation. After removal of the plasma, the red cells are usually suspended in an additive solution, which is a crystalloid solution allowing for storage for up to 42 days at refrigerator temperatures of 1-6°C. The mass of red cells in a red cell concentrate varies between 150-250 mls, but on average is about 200 mls. This product also contains the additive solution, which has a fixed volume of 100 mls, and a small amount of "carry over plasma" (25-50 mls), such that the actual volume of the red cell concentrate is between 280-400 mls. The hematocrit is 50-60. These characteristics are shown in Table 27.1.

The indications for red cell transfusion are best divided into actively bleeding patients and those with normovolemic anemia. Patients who are actively bleeding, as in trauma, surgery or spontaneous bleeding from the gastrointestinal tract, may be candidates for red cell transfusion. The initial approach in these patients is to transfuse a crystalloid solution, such as saline, rather than red blood cells, but at a critical point if the bleeding is excessive, and particularly if the patient is known to be anemic prior to bleeding, red cell transfusion may be appropriate. The purpose of the blood transfusion in this context is to restore intravascular volume and also allow the delivery of oxygen to tissues. The dose (number of units) of red cell transfusion in acutely bleeding patients is determined by the treating physician based on the extent of the hemorrhage. Laboratory values such as hemoglobin and hematocrit, even when available, may not be useful, and clinical parameters such as vital signs and estimates of acute blood loss expressed in blood volumes are more important. Guidelines for red blood cell transfusion in acute blood loss are shown in Table 27.2.

The second situation in which red cell transfusions are administered is the clinical situation known as normovolemic anemia. Normovolemic anemia is a situation in which patients have a low hemoglobin, are hemodynamically stable, and in whom there is no imminent expectation of acute blood loss. Although, by definition, anemia occurs if the hemoglobin decreases below 12.5 g/dl, in practice normovolemic anemia often refers to patients with a hemoglobin of 10 g/dl or less. There is considerable controversy surrounding the level of hemoglobin at which red cell transfusion may be appropriate (Trigger, Chapter 26), but, in general, patients with hemoglobins less than 7 g/dl, particularly older patients, may experience clinical symptoms consistent with insufficient oxygen delivery. Virtually all the controversy exists, therefore, regarding transfusing red cells to patients

Clinical Transfusion Medicine, by Joseph D. Sweeney and Yvonne Rizk. © 1999 Landes Bioscience

Table 27.1. Red blood cell transfusions

<u>Product:</u>	Red Blood Cells (Packed Cells)	
<u>Characteristics:</u>	Volume:	280-400 ml
		150-250 ml RBC
		2×10^9 white cells
		Hct 50-65

<u>Pharmacological Effect:</u>   Improve O_2 carriage and delivery

<u>Indication:</u>

I.   Acute Bleeding →   • Replaces volume
     (Hypovolemia)   • Improves oxygenation

II. Anemia   →   • Improves oxygenation
     (Normovolemic)

          Anemia (Hb 7-10 g/Dl) with
          Symptoms of impaired oxygenation

<u>Dosage:</u>   1 Unit per 70 Kg per 1 g increase in Hb

27

Table 27.2. Classification of acute hemorrhage and recommendations regarding red cell transfusion

	Class I	Class II	Class III	Class IV
Percent Loss of Blood Volume	0-15%	15-30%	30-40%	> 40%
Approximate Volume Loss (Adult)	< 750 ml	750-1500 ml	1500-2000 ml	> 2000 ml
Vital Signs	mild tachycardia	tachycardia; decrease pulse pressure; tachypnea	tachycardia; tachypnea; hypotension	tachycardia; unmeasurable blood pressure
Replacement Fluids	saline 1-2 liters	saline initially. possible red cell transfusion	saline; probably red cell transfusion	red cell transfusion required

(Advanced Trauma Life Support Subcommittee, American College of Surgeons).

with normovolemic anemia and hemoglobins between 7-10 g/dl and practices vary greatly between individual physicians, even within the same institution.

Chapter 26 outlines the principles regarding clinical decision making in normovolemic anemia, and applications of these principles to individual patients will reduce inappropriate decision making regarding transfusion. The decision to transfuse is based on the degree of anemia in relation to clinical circumstances. It is helpful to document the rationale for the transfusion of red cells in the patient's record such as "Hemoglobin 8.5 g/dl; patient clinical symptomatic with fatigue at rest: one unit of red cells to be transfused with posttransfusion monitoring of the hemoglobin (between 1-24 hours)".

Although dosing of red cells in patients with acute blood loss is guided entirely by the extent of bleeding, in normovolemic anemia it is important to consider two factors: (1) the desired increase in hematocrit and (2) patient's intravascular volume.

This is illustrated in the formula given in Figure 27.1, which shows that the volume of red cells to be transfused (in ml) is equal to the desired difference in hematocrit posttransfusion multiplied by the blood volume of the recipient. In practice, this means that for any given desired increase in hematocrit, patients with a larger intravascular volume will acquire a higher dose (more units) than those with a smaller intravascular volume. The clinical application of this principle is that elderly low weight females may benefit adequately from a single unit of red blood cells, whereas larger males will generally require higher doses. Whenever the hematocrit is in a borderline range (7-10 g%), it is also acceptable to transfuse a single unit of red cells and observe for a clinical response and measure the laboratory response.

Example 1: A 50 Kg 80-year-old female with intermittent chest pain has a hematocrit of 24 (0.24). The desired posttransfusion hematocrit is 30. What dose of red cells is required? How many units?

27

General Formula for calculating the dose of red cells is as follows:

If, Hct^F = Desired posttransfusion hematocrit (Fraction e.g., 0.30)
Hct^i = Pretransfusion (initial) hematocrit (Fraction e.g., 0.21)
BV = Blood volume of recipient (ml)
RUV = Volume of red blood cells in the unit (ml)

$$Hct^F = \frac{Hct^i \times BV + RUV}{BV}$$

then, $Hct^F \times BV = Hct^i \times BV + RUV$

$Hct^F \times BV - Hct^i \times BV + RUV$

or $BV (Hct^F - Hct^i) = RUV$ $\quad$ # units = $\frac{BV (Hct^F - Hct^i)}{200}$

i.e., volume of red cells is determined by the Hct difference *multiplied* by the blood volume.
Assume: BV = 70 ml/Kg and 1 unit = 200 ml of red blood cells.

Fig. 27.1. Calculation of a dose for red blood cells, expressed as ml of packed cells or "Units" of red blood cells.

Blood volume (BV) = 50 x 70 mls = 3,500 mls

Pretransfusion Hct (HCT^I) = 0.24

Posttransfusion Hct (HCT^F) = 0.30

Then: RBC (mls) = 3500 (0.30-0.24)

= 3500 (0.06)

= 210 mls

Therefore: The dose is 1 unit.

Example 2:   A 75 Kg 68-year-old male with intermittent chest pain has a hematocrit of 24 (0.24). The desired posttransfusion hematocrit is 30 (0.3). What dose of red blood is required? How many units?

Blood volume (BV) = 75 x 75 mls = 5625 mls

Pretransfusion Hct (HCT^I) = 0.24

Posttransfusion Hct (HCT^F) = 0.30

Then: RBC (mls) = 5625 (0.30-0.24)

= 5625 (0.06)

= 338 mls

Therefore: The dose is 2 units.

Newer red cell products will soon be available. Recently, a larger blood collection (500 ± 10% versus 450 ± 10%) has been approved and thus the average volume (mass) of red cells per unit may increase to 220 mls. This potentially will reduce the dosage as expressed in units. In addition, new apheresis devices now allow the collection of "two units" of red cells from a donor. This can yield a dose from 180 mls to over 400 mls per donation. These new developments indicate that traditional dosing based on "units of red cells" will soon be obsolete.

27

Platelets: Indications and Dosing

Blood platelets are currently manufactured in one of two ways. Whole blood donors may donate a unit of blood from which a platelet concentrate is manufactured. In this process, the unit of blood is subjected to two centrifugational steps. The first step is called a soft spin, which makes platelet rich plasma and a concentrated (packed) red cell. The platelet rich plasma is expressed from the bag and then subjected to a second centrifugation called a hard spin, after which the platelets are concentrated into a small amount of plasma (35-60 mls). In some European countries, the centrifugation is reversed, and the platelets are manufactured from the layer between the red cells and plasma, called the buffy coat. Either way, the end product is called a unit of platelets or a random donor platelet unit.

Alternatively, donors may have their blood anticoagulated and drawn into special machines, called apheresis machines. In this procedure, platelets are separated by centrifugation, and the red cells returned to the blood donor together with most of the plasma. This procedure takes 50-90 minutes. The correct name for this product is platelet pheresis, but is more commonly known as single donor platelets or apheresis platelets. Platelet pheresis, or single donor platelets, have a higher content of platelets (absolute number, yield or potency) than are present in a unit of platelets (random donor platelets) derived from a whole blood donation. Approximately 5-8 random donor units of platelets are equivalent to one apheresis product. The characteristics of platelet products are shown in Table 28.1.

The clinical indications for platelet transfusions are to prevent or stop bleeding in patients with low platelet counts (thrombocytopenia) or less commonly, in patients with dysfunctional platelets (thrombocytopathy). These indications occur in several different types of clinical settings. First, patients with severe thrombocytopenia. The most common indication in this setting is to prevent spontaneous bleeding, particularly spontaneous intracranial bleeding. Most current literature now shows that this is unlikely to occur unless the platelet count decreases below 10×10^9/L (10,000/mm^3) and a high risk is not present until the platelet count decreases below 5×10^9/L (5,000/mm^3). In the past, a threshold value of 20×10^9/L (20,000/mm^3) was commonly used by hematologists to prevent spontaneous bleeding in patients with acute leukemia and bone marrow transplantation, but this is now obsolete. The second clinical situation is thrombocytopenia in a patient for whom an invasive diagnostic procedure is imminent, such as liver biopsy, colonoscopy with biopsy, bronchoscopy with biopsy, etc. The transfusion trigger platelet count is unknown, but is commonly considered to be 50×10^9/L or lower. Patients with platelet counts below 50×10^9/L, may, therefore, be appropriate candidates for prophylactic platelet transfusions in this setting, although many such procedures can be performed without platelet transfusion, depending on the

Clinical Transfusion Medicine, by Joseph D. Sweeney and Yvonne Rizk. © 1999 Landes Bioscience

---

Table 28.1. Platelets

Product:

Human platelets suspended in plasma. The platelets comprise only 2-4% of the total volume; the remainder, 96-98% is plasma.

Characteristics:

	Whole Blood Donation	Apheresis Donation
Potency	$5 - 8 \times 10^{10}$	40×10^{10}
Volume (ml)	35 - 60	180 - 400
Labeling	Platelets	Platelets, pheresis
Common Usage	Random Donor Units	Single Donor Unit

Pharmacological Effect:

Increase the Platelet Count and Prevent or Stop Bleeding

Indications:
1. *Thrombocytopenia ($5 - 20 \times 10^9$/L) to prevent spontaneous bleeding
2. *Thrombocytopenia ($< 50 \times 10^9$/L) with active bleeding or prior to invasive procedure
3. Normal platelet count: Qualitative (abnormal) platelet function

Doses:
1 unit/10 Kg weight; 4 units/m2 Surface Area
1 Platelets, Apheresis

(* 20×10^9/L = 20,000/mm^3)

28

skill of the operator. A third clinical situation is the presence of thrombocytopenia in a patient prior to a surgical procedure. In this situation, the underlying cause of the thrombocytopenia and the nature of the surgical procedures are important. A preoperative trigger count of 50×10^9/L (50,000/mm^3) is often used but a lower trigger may be appropriate. Considerations are whether the procedure in itself is ordinarily associated with excessive blood loss; whether the bleeding can be well visualized and controlled by local surgical measures; or whether small amounts of bleeding in a closed space would create a residual functional problem for the patient, for example, neurosurgical procedures or ophthalmic surgery. In these latter situations, a preoperative trigger of $80-100 \times 10^9$/L (80,000-100,000/mm^3) is sometimes advocated for such surgery. A fourth clinical situation is a thrombocytopenic ($< 100 \times 10^9$/L or 100,000/mm^3) patient who is actively bleeding, for example, acute gastrointestinal bleeding. There is very little data available to guide platelet transfusion in this context. The concern is that the low platelet count could be either a significant contributing cause to or exacerbation of the degree of blood loss. In this setting, it is probably wise to treat if the platelet count is less than 50×10^9/L and, possibly, if the platelet count is less than 100×10^9/L. If large volumes of red cells are transfused, platelet transfusion will certainly be required on account of hemodilution and may need to be repeated. A fifth situation arises when the platelet count is normal but the platelets are considered to be dysfunctional, such as in a patient with excessive chest tube drainage after cardiopulmonary bypass (for example, in excess of 300 ml/hour). Empiriplatelet transfusions may be appropriate in these patients and useful in avoiding a surgical re-exploration. A common situation is the patient with a normal platelet count who has taken aspirin and or similar drugs and requires a surgical procedure or an invasive diagnostic procedure. Deferral of the procedure for 48-72 hours is optimal since platelet function will return to normal if aspirin is discontinued for this time. For other nonsteroidal drugs, 6-8 hours may be adequate, since the effect is reversible more quickly. This is because aspirin irreversibly acetylates an enzyme, cyclo-oxygenase, in the platelet and 2-3 days are required for the bone marrow to produce 20-30% normal (nonacetylated) platelets. Other nonsteroidal drugs reversibly inhibit this enzyme, and the effect disappears when the drug has been cleared. Ticlopidine (Ticlid®) and Clopidogrel (Plavix®) have a different mechanism of action and discontinuation of these drugs for at least 10 days is needed in order to reverse the antiplatelet effect. If an urgent surgical procedure is required, it is best to have platelets available for possible transfusion and if microvascular oozing is observed intraoperatively, transfusion may be appropriate. In neurosurgery or ophthalmic surgery, however, where minimal amounts of excessive blood loss could cause severe functional problems, prophylactic platelet transfusion may be appropriate before surgery. The dose of platelets needed to reverse an aspirin effect is known to be less than "standard dose" since as few as 15-20% of nonaspirinized platelets will suffice. The dose, therefore, should not normally exceed four units.

Platelet dosing is very controversial and there is no such thing as a "standard platelet dose". Surveys of different institutions indicate that between 5-10 units of platelets or equivalent is fairly routinely administered per transfusion. The generally recommended dose is 1 unit of platelets per 10 kg body weight or 4 units/m² surface area. If platelet pheresis is available, the dose is the content of the single donor product. As with red cells, it is useful to consider the clinical situation, the pretransfusion platelet count, the desired posttransfusion platelet count and the size of the intravascular volume of the recipient (body weight). A generally suggested dose of random donor platelets might be five units. Higher doses of platelets have traditionally been transfused, such as 8-10 units, but this may have arisen because of less attention to quality control in manufacturing of platelets in the past and may have resulted in lower quality products (i.e., lower platelet content). Increasing the number of units, therefore, was to compensate for this uncertainty and increase the likelihood of an adequate response. This is no longer the situation in most Blood Centers. The dose used should have a reasonable expectation of success, i.e., absolute increase in the platelet count of $20-40 \times 10^9/L$. A suggested platelet algorithm for adult dosing is shown in Figure 28.1.

New developments in blood collection technology point to the increasing use of apheresis machines for the collection of all blood components. If this is the case, a 'standard apheresis product' could become the only product available in the future.

One of the more complicated problems encountered in clinical practice is the management of patients refractory to platelet transfusions (Table 28.2). These patients are typically cancer patients or bone marrow transplant patients receiving

Table 28.2. Causes of refractoriness to platelet transfusion and management strategies

1. Antibiotic agents (especially vancomycin or cephalosporins). Considering discontinuing or changing these drugs.

2. Amphotericin B: Evaluate for continuing need.

3. ABO incompatible platelets: Transfuse ABO identical platelets and monitor the response in platelet increase of 10-60 minutes.

4. Hypersplenism: Consider lowering the trigger for transfusion.

5. Fresh platelets (less than 36 hours old) may provide better platelet increases.

6. HLA alloantibodies: Consider HLA selected (matched) platelets or crossmatched platelets, if available.

7. Platelet-specific antibodies: Consider crossmatched platelets, if available.

8. For all patients: Consider transfusing red cells to maintain a minimum Hct of 32–35.

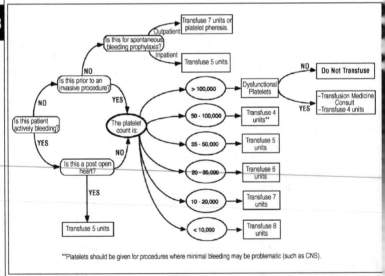

**Platelets should be given for procedures where minimal bleeding may be problematic (such as CNS).

Fig. 28.1. Suggested algorithm for initial adult platelet dosing in different clinical situations. This dose assumes a 70 Kg recipient and blood volume of 5 liters. Subsequent doses should be determined by the clinical circumstances.

platelet transfusions as prophylaxis for spontaneous bleeding. In this situation, little or no increase in the platelet count is observed after the transfusion of a platelet product and an actual decrease may sometimes be observed! This is a perplexing problem for both the Blood Bank and the treating physician. There are many causes of this problem, but in some recipients the refractoriness is due to alloantibodies against class I HLA antigens or platelet specific antigens (immune case). In assessing these patients, nonimmune causes should be sought such as the use of antibiotics and antifungals especially vancomycin and amphotericin, or splenomegaly. ABO incompatibility should be considered; e.g., transfusing A or B platelets to an O recipient. HLA selected platelets should only be requested after these have been evaluated. It has also been suggested that these patients respond better to fresh (less than 36 hours) platelets, when available. In many instances, the response to HLA selected (matched) platelets is disappointing. These patients should be managed by transfusing at least one dose of platelets daily in order to meet the "endothelial" need for platelets; the actual increment observed need be of less concern. This concept of "endothelial" need is that a small percentage (7%) of the platelet mass is consumed in normal (healthy) subjects daily in sealing breaks in endothelial integrity. In thrombocytopenic patients, this same mass (or more) of platelets is still required, and must be supplied by allogeneic platelets since the autologous platelets mass is inadequate. In addition, red blood cells should be transfused to maintain a hematocrit at between 32-35. This reduces plasma volume thereby effectively increasing the concentration of platelets. There is data that this approach improves one surrogate test of platelet function, the bleeding time. The practice of transfusing massive doses or multiple daily doses of platelets to these patients is wasteful, does not have any empiric justification, and should be resisted.

28

Plasma and Cryoprecipitate: Indications and Dosing

Plasma, and a product derived from plasma called cryoprecipitate, are sometimes called acellular components, since they lack viable cells. These products have different indications and need to be discussed separately.

PLASMA

The most common plasma product transfused is known as fresh frozen plasma (FFP). FFP is plasma which has been separated from whole blood and frozen within 8 hours of collection. Many blood centers manufacture a product similar to fresh frozen plasma, but which is frozen within 24 hours. For practical purposes, these products should be considered interchangeable and are a good source of both stable and labile blood coagulation factors. "Liquid plasma" is plasma which is separated from the red blood cells and has never been subjected to freezing. It is a reasonable source of the stable clotting factors (II, VI, IX and X), but the instability of the labile factors (FV and FVIII) has made this product unpopular.

Frozen plasma contains a volume of approximately 220 ml and essentially all plasma proteins. The characteristics of FFP are shown in Table 29.1, together with the major indications for use. The majority of plasma is transfused in order to replace blood-clotting factors in patients who are either actively bleeding (spontaneous or surgery) or in patients with prolonged clotting times prior to an invasive procedure. The most common situations where this is encountered is in patients with liver disease with a prolonged prothrombin time (PT); in bleeding patients who are vitamin K deficient or known to be taking an oral anticoagulant; in patients massively transfused in surgery or trauma; or in patients with disseminated intravascular bleeding who are actively bleeding. Only a small amount of total plasma is used in the treatment of the hereditary bleeding disorders (factor V or factor XI deficiency), for which either plasma derived or recombinant concentrates are not currently available.

The appropriateness of plasma transfusion is one of the more controversial areas in clinical transfusion. This is particularly the case regarding the decision to transfuse plasma to patients with a mild prolongation of the prothrombin time prior to diagnostic procedures such as liver biopsy, paracentesis, or lumbar puncture, or prior to surgical procedures in which blood loss is ordinarily minimal (i.e., rarely transfused with red cells). A common misconception is that patients will bleed excessively if a mild prolongation of the prothrombin time is present. This is not substantiated by available data, which suggests that the likelihood of

Clinical Transfusion Medicine, by Joseph D. Sweeney and Yvonne Rizk. © 1999 Landes Bioscience

---

Table 29.1. Plasma

Product
   Anticoagulated plasma in frozen state

Characteristics
   Volume: 200-225 ml (= 1 unit)
   Contains **all** plasma proteins

Pharmacological Effect
   Increase plasma clotting factors and prevent or stop bleeding

Indications
   1. Acquired bleeding disorders with active bleeding or prior to an invasive procedure. (liver disease; vitamin K deficiency or warfarin; disseminated intravascular coagulation; dilutional coagulopathy).

   2. Hereditary bleeding disorders, when a concentrate is not available (FV or FXI deficiency).

Dose:
   10 - 15 ml/Kg

---

bleeding only occurs with more marked prolongations of the PT, i.e., in excess of 1.5 times the mean value of a control-normal population, often corresponding to a PT of 18 sec. or greater as discussed in Chapter 26.

Active bleeding in patients with a prolonged clotting time constitutes a reasonable indication for plasma, regardless of the degree of prolongation particularly if the acute blood loss is being managed with red cell transfusion. In surgical patients, however, it is important to search for an anatomic cause of the bleeding and have this corrected.

A different clinical situation arises in a patient with a prolonged PT who requires a surgical procedure, in which a large blood loss may occur intraoperatively. In such patients, dilutional coagulopathy will likely occur early after fewer units of red cells have been transfused (i.e., 0.3-0.6 blood volumes), resulting in significant microvascular bleeding than in a patient who is hemostatically competent preoperatively. Prophylactic administration of plasma early in the surgical procedure (after 2-4 units of red cells) may constitute appropriate judgment, since it may avert the occurrence of the dilutional coagulopathy.

Patients who are hemostatically competent (normal prothrombin time) preoperatively may also require plasma intraoperatively in certain procedures in which the blood loss is excessive (> 0.5 blood volume), for example, spinal surgery, extensive oncologic surgery with massive blood loss, or vascular reconstructive surgery. In the past, it was considered that platelet transfusions were important early in the management of such patients (Chapter 14). Much of the data supporting this, however, was from an era in which the red cell products transfused were suspended in plasma (prior to 1983). After massive transfusions of such red cell products, these patients had already received significant amounts of replacement plasma containing coagulation factors, particularly the stable factors (fibrinogen, FVII, FIX, FX). Currently, the product most commonly transfused is red cells suspended in a crystalloid solution (Chapter 27). In addition, the use of intraoperative salvage will result in the return of autologous red cells suspended in saline, in which both clotting factors and platelets are absent. Available data suggests that a blood loss corresponding to as little as 0.5 blood volumes (arbitrarily 5-6 units in a standard weight individual) *may* be associated with clinically evident microvascular oozing, which responds clinically to the transfusion of plasma.

Plasma dosing is often inappropriate on account of the practice of prescribing plasma as "units" desired. It is not uncommon to observe a request for either 1 or 2 units of FFP for an adult. This constitutes a volume of between 200-450 ml and will not achieve a significant increase in clotting factors. The appropriate dose is at least 10 ml/Kg, and doses as high as 20 ml/Kg may be appropriate in patients with continuing active bleeding and dilutional coagulopathy. These doses will increase blood clotting factor levels to at least 25-33% of normal levels, which is considered adequate for hemostasis. Clearly, further doses of plasma will be required if blood loss continues and is replaced with allogeneic red cells in crystalloid solution or salvaged autologous red cells suspended in saline. Since a unit of plasma can be considered roughly to have a volume of approximately 220 mls, the volume of plasma required in ml/Kg can simply be divided by 200 to achieve the desired number of units. For a 70 Kg subject, therefore, a minimal dose would be 700 mls or 3-4 units of FFP and a dose of 15 ml/Kg would correspond to 1,000 ml or about 5 units. It is apparent from these calculations that a request for 1 or 2 units of FFP represents underdosing.

One inappropriate practice is the routine transfusion of a unit of FFP on a formula basis for every 2-4 units of allogeneic red cells transfused in an otherwise hemostatically competent patient (no history of bleeding; normal PT). This practice has no empiric clinical justification, but it is likely to have evolved as a preventative measure for dilutional coagulopathy. This practice, however, results in excessive use of plasma, since most patients undergoing operative procedures requiring 2-4 units of red blood cells will not develop a dilution coagulopathy. It is preferable to wait until a large blood loss has occurred (0.5-1.0 blood volume), observe for clinical evidence of microvascular oozing and treat with the appropriate dose (10-15 ml/Kg).

Two newer plasma products have recently become available. Solvent detergent plasma (SD plasma) is plasma produced from a pool of donations (about 2,500). The pool is subjected to viral inactivation by a process called solvent-detergent (SD) treatment. This process inactivates some hepatitis viruses (hepatitis B and C) and HIV-1. It does not inactivate all viruses, however. Although potentially safer because of the viral inactivation step, the larger pool of donors is of concern since it potentially creates a scenario in which rapid spread of unknown viruses which are not inactivated by the SD process could occur. A second product is fresh frozen plasma; donor retested (FFP-DR). This involves quarantining the frozen plasma from a donation for 90-120 days until the donor returns to donate. If all infectious disease testing is satisfactory (normal) at the subsequent donation, the frozen plasma from the previous donation is released from quarantine. This plasma is likely to have a reduced risk of viral disease transmission (Chapter 34).

Cryoprecipitate is a product derived from the slow thawing of frozen plasma and is routinely produced in Community Blood Centers. It differs from plasma in that it contains predominantly high molecular weight glycoproteins, such as fibrinogen, factor VIII, von Willebrand factor and factor XIII. About 50% of the original amount of these proteins are concentrated in a small volume (5-15 ml). Table 29.2 shows the characteristics and clinical indications for cryoprecipitate. Overall, the most accepted current indication for cryoprecipitate is the treatment of a bleeding patient with hypofibrogenemia and general agreement that a fibrinogen level of less than 100 mg/dL is a reasonable trigger. This level is most frequently seen in severe disseminated intravascular coagulation or in dilutional coagulopathy. In some surgical settings, higher postoperative fibrinogen triggers are used, such as active bleeding in cardiac patients or in patients with hepatic resections (150-200 mg/dL). This is on account of concern that a further precipitous reduction in fibrinogen could occur in these patients, which would exacerbate clinical bleeding. An additional indication for the use of cryoprecipitate is the treatment of uremic bleeding (Chapter 18). Cryoprecipitate has also been transfused to patients with hereditary platelet defects either prior to an invasive procedure or where there is active bleeding, as it is known to shorten the bleeding time in these patients. The use of cryoprecipitate in hemophilia A (factor VIII deficiency) or von Willebrand's disease is now uncommon, as more appropriate treatment regimens are available (Chapter 21). Severe factor XIII deficiency (< 1% factor XIII) is an exceedingly rare disorder for which there is no concentrate available in the United States. This is best managed by the administration of cryoprecipitate once or twice monthly, since factor XIII has a long half life (14 days).

29

Table 29.2. Cryoprecipitate

Product:
Anticoagulated product containing cryoprecipitated proteins

Characteristics:
Volume: 5 - 15 ml (= 1 Unit; 1 BAG)
Contains high molecular weight glycoproteins such as fibrinogen (300 mg/unit); Factor VIII (80-100 U/unit); von Willebrand factor; factor XIII

Pharmacological Effect:
Increase plasma levels of high molecular weight clotting factors

Indications:
1. Hypofibrinogenemia: Fibrinogen < 100 mg/dL with active bleeding or fibrinogen < 200 mg/dL in a postoperative patient with excessive bleeding.
2. Uremia or hereditary platelet disorder
3. Factor XIII Deficiency

Dose:
1 unit/10 Kg wt; Frequently 10 BAGS

Dosing of cryoprecipitate tends to be unscientific. In general, 10 units (or 10 bags) of cryoprecipitate will increase the level of fibrinogen by 80-100 mg/dL in an average person. This may, however, be short lived and further doses may be required. In the treatment of uremic bleeding, the dose has been standardized empirically to 10 units, irrespective of body weight or the degree of uremic dysfunction. For pediatric patients, a dose of 1-2 U/Kg is reasonable. This weight-based dosing could also be applied to adults, but, in practice, only minor savings in cryoprecipitate use would occur.

29

Leukocytes: Indications and Dosage

Leukocytes are probably the least frequent blood product requested of a transfusion service. There are a variety of leukocyte products, some of which are mostly of research interest. For example, there has been interest in the use of interleukin-2 stimulated lymphokine activated killer cells (IL2-LAK) and in the use of ex vivo monocytes stimulated with gamma interferon [EVLA] treatment in the adoptive immunotherapy of cancer. Both of these are largely experimental and have not come into routine use at this time. This Chapter will focus exclusively on granulocyte concentrates.

The types of granulocyte concentrates available are shown in Table 30.1. As discussed with platelet products (Chapter 28), granulocyte concentrates can be produced from either a whole blood donation, in which case it is known as a buffy coat, or by the use of apheresis devices. Apheresis granulocytes are essential for adult recipients and may be the preferred product for neonates, but timely availability can limit their use for this latter population.

The buffy coat product has a low volume, similar to random donor platelets but, unlike platelets, contains large numbers of red cells and thus requires ABO compatibility. Approximately 65-75% of the leukocytes present in the whole blood donation are concentrated in the buffy coat and the content, therefore, of granulocytes is approximately 1×10^9. The apheresis granulocyte concentrate has a much larger volume. It will contain many red blood cells and have a hematocrit of approximately 20, therefore, also requiring ABO compatibility. The granulocyte content in the standard apheresis granulocyte concentrate is generally between $1-3 \times 10^{10}$ i.e., approximately 10 times as many granulocytes as a buffy coat product. Recently, there has been interest in stimulating normal healthy donors with GCSF prior to white cell collection. The white cell count of the apheresis donors increases to 20×10^9/L (20,000 mm^3) or greater and granulocyte content of the granulocyte concentrate collected is correspondingly greater, containing up to 10×10^{10} granulocytes. These products are neither licensed nor generally available as yet.

The indications for the use of these concentrates are shown in Table 30.2. For practical purposes, buffy coats are used almost exclusively in the treatment of neonatal sepsis with neutropenia or qualitative granulocyte disorders. In adult practice, apheresis granulocytes are used in infected neutropenic adults. Currently available granulocyte concentrates are not known to be useful in the prophylaxis of neutropenic infections and only patients with active infections under conditions as suggested in Table 30.2, may be candidates for granulocyte concentrates. Most oncologists treating patients with neutropenic fever have abandoned the use of granulocyte concentrates as clinical results have been disappointing.

30

Table 30.1. Granulocytes concentrates

1. Type of Products:
   a) Buffy coat from whole blood donations
      i) Buffy coat: volume 30-50 ml
         - 8-15% of total RBC
         - 65-75 % of total leukocytes (1×10^9)

   b) Apheresis Granulocytes:
      - Volume: 200 ml
      - Hct of approximately 20
      - Standard product: $1\text{-}3 \times 10^{10}$ granulocytes
        GCSF stimulated donors: $4\text{-}10 \times 10^{10}$ granulocytes

2. Granulocytes should be transfused as soon as possible after collection and never transfused using a leukoreduction filter. This may preclude completion of the usual tests for viral markers and justification of the clinical need may be required in writing from the prescribing physician.

3. Commonly, granulocytes are irradiated (Chapter 37) since the recipient is frequently immunocompromised: Some centers *routinely* irradiate all granulocyte concentrates.

4. These products may need to be manufactured from donations which are cytomegalovirus seronegative, as a leukoreduction filter cannot be used.

5. Transfuse over two hours; reactions are not uncommon and managed by slowing infusion, steroids, acetaminophen or antihistamine as appropriate. Amphotericin B infusion should not be concurrent with granulocyte transfusions.

30

Table 30.2. Granulocytes: indication and dosage

1. a) Buffy coat: Neonatal sepsis with neutropenia

   b) Apheresis Granulocytes:
      Suspected or proven gram negative sepsis or fungal infection
      i) With evidence of persisting infection, e.g., fever > 38.5°C x 48 h despite treatment with appropriate multiple antibiotics using an appropriate dosage regimen.

         and

      ii) Neutrophil count < 0.5×10^9/L, without expectation of white cell recovery for 5 days

         and

      iii) Adult with an expected survival of > 2 months

2. Dosage and Scheduling:
   a) Buffy coat—1 product

   b) Apheresis Granulocytes—1 product QD x 3-5 days

It is possible, however, that granulocyte concentrates will prove of greater benefit in the future if data on collections from GCSF stimulated donors show better clinical responses and assuming that GCSF does not cause significant side effects in the healthy donor. Early experience with these higher potency granulocyte products is that recipients show increases in white cell count posttransfusion (which is not observed with the standard granulocyte concentrates) and resolution of fever. Thus, the standard granulocyte concentrates appear limited by a lack of potency, and therapeutic efficacy may only be achieved in most recipients when this is improved.

With regard to scheduling of these products, buffy coats are usually transfused as a single dose, but multiple doses may be administered on a daily basis. For apheresis products, however, it is common practice to transfuse a product on a daily basis for a total of 3-5 days. Although many patients receiving granulocytes are also receiving leukocyte-reduced blood products, leukoreduction filters (Chapter 36) must never be used for granulocytes. On account of this, granulocytes may need to be from CMV seronegative donors, if such an indication exists in the recipient.

Most transfusion services or blood centers will ensure that the product is irradiated either prior to shipment or transfusion (Chapter 37). The absolute need for irradiation of all granulocytes is not established, but if doubt exists, it is best to irradiate as many recipients of these products are immunocompromised.

Granulocytes are stored at 20-24°C without agitation and the shelf life of granulocyte concentrates is 24 hours. It is recommended, however, that they be transfused promptly and within 8 hours of collection, if possible. This will require the Blood Center to ship these products often without completion of viral disease testing and therefore documentation of the urgent clinical need will be needed. Granulocytes are transfused slowly (2 hours) and as emphasized, leukoreduction filters must never be used; the standard nylon meshwork filter (Chapter 8) is used in the administration set. Reactions to granulocytes are common, but most can be managed with acetaminophen, steroids or antihistamines. Severe reactions causing pulmonary edema and acute dyspnea are most feared and will need careful monitoring and intervention with ventilation if these occur. Some of these recipients may be receiving amphotericin B as an antifungal agent and it is recommended practice that the granulocyte transfusion and the amphotericin B infusion be separated by several hours in order to prevent pulmonary reactions.

30

Blood Derivatives:
Indications and Dosage

The term blood derivatives refers to a family of blood products which are derived from a pool consisting of many thousands of blood donations. Only the plasma components of the whole blood or apheresis donation are used in these pools.

The major characteristics of the blood derivatives are shown in Table 31.1. Currently available blood derivatives *do not* have a blood type label, and therefore, ABO compatibility and/or Rhesus compatibility are not relevant for transfusion purposes. Unlike most blood components, which are manufactured in community blood centers, blood derivatives are manufactured in large (commercial) fractionation plants and the end-product may be in liquid or lyophilized form. It is important to appreciate that all blood derivatives are now subjected to multiple viral attenuation processing steps during manufacture. These processes may be physical and or chemical and of proven efficacy in inactivating or destroying viruses. Virus disease transmission (Chapter 34), is, therefore, much less of a consideration than in the past. Regardless, recombinant products are now available for factor VIII and factor IX deficient patients, which is greatly reducing the need for plasma derived products.

There has been an extensive clinical experience with albumin and it has never been implicated in the transmission of virus diseases. Albumin has traditionally been virally attenuated using a pasteurization process (60°C for 11 hours). This appears very effective in destroying virus in the pool. Albumin is available either as a 5% (5 g/dL) or a 25% (25 g/dL) salt free product. Each formulation contains the equivalent amount of albumin present in one unit of plasma; the 25% solution has low electrolytes. The 25% solution should not be dissolved in sterile water, if the 5% solution is desired, as the resulting solution is hypotonic and may cause hemolysis. The volume of each vial of albumin is 250 ml for the 5% solution and 50 ml for the 25% solution. The indications for the use of albumin are not well defined in clinical practice. General situations where albumin has been used are in hypovolemic states associated with hypoalbuminemia or in promoting salt loss in association with diuresis in nephrotic patients. Albumin has also been used to prevent hypotension when large volumes of third space fluid have been removed. Less expensive colloidal preparations are available, such as Hetastarch (Hespan) which for many patients may be an acceptable alternative. Plasma protein fraction (PPF) is very similar to albumin. It is supplied only as a 5% solution and has a volume of 250 ml. Plasma protein fraction differs from albumin only in the β-globulin content (PPF has a slightly higher β-globulin content). It is questionable whether any real difference of clinical importance exists between 5% albumin and PPF.

Clinical Transfusion Medicine, by Joseph D. Sweeney and Yvonne Rizk. © 1999 Landes Bioscience

Table 31.1. Characteristics of the plasma derivatives

1. Plasma proteins manufactured from pools containing 5,000-20,000 donations.

2. Do not have an ABO blood type label.

3. Manufactured in a fractionation plant.

4. Treated using chemical or physical methods which inactivate viruses and bacteria.

5. Types of products available:

> Albumin (5% or 25%); Plasma protein faction
> Immunoglobulins: IVIG, anti-D (Rhogam; Win-Rho)
> Coagulation Factors (VIII; IX)
> Protein Inhibitors (Antithrombin III; Antitrypsin)

Table 31.2. Therapeutic uses of plasma derivatives

1. Albumin/PPF: Increase oncotic pressure and reduce edema.

2. Intravenous Gamma Globulin (IVGG)

   a) Increase immunoglobulins and prevent or treat infections.
   b) Immunomodulate the immune system in autoimmune diseases.

3. Clotting Factors:
   Prevent or treat bleeding in Hemophilia A or B.

4. Antithrombin III: Prevent venous thrombosis postpartum in hereditary ATIII deficiency; treatment of venous thrombosis in ATIII deficiency.

5. Antitrypsin: Prevent pulmonary emphysema and hepatic cirrhosis in hereditary α-1-AT deficiency.

31

The immunoglobulin products currently available for intravenous use represent a significant advance in the treatment of many diseases. Intravenous immunoglobulins are used in two situations: (a) to increase levels of immunoglobulins in patients with hypogammaglobulinemia: either congenital hypogammaglobulinemia, such as in children, or acquired hypogammaglobulinemia, such as in chronic lymphocytic leukemia. The dosage used is approximately 0.1 g/Kg, i.v. at intervals of 2-4 weeks; (b) immunoglobulins are also used as immunomodulatory agents. Important uses are the treatment of idiopathic thrombocytopenic purpura (ITP); autoantibodies to factor VIII, or, more recently, acute Guillain-Barré syndrome. Less well accepted indications are the management of thrombocytopenic patients refractory to platelet transfusions or in the treatment of patients with warm autoimmune hemolytic anemia, not responding to steroids. The dose for these conditions is 2 g/Kg either daily for 5 days or 1 g/Kg on two alternate days. Anti-D (Win-Rho) in higher doses than those used as prophylaxis in obstetrics (Chapter 21), has recently been used to treat ITP in Rhesus (D) positive patients. Doses of anti-D are 50-75 µg/Kg, which are 10-15 times the dose commonly administered as prophylaxis in obstetrics (300 µg). Hemolysis is common, but is rarely a clinical problem. The clinical response to Win-Rho in ITP appears equivalent to IVGG and the cost of Win-Rho is less.

Clotting factor concentrates derived from plasma are predominantly factor VIII and factor IX. These products are still available and in use (1999), but are largely being replaced by recombinant protein products. The doses used are as indicated in Chapter 21. Because of the availability of recombinant products, it is important to seek the advice of an experienced hematologist prior to using any of these plasma derived products at the present time.

Other plasma derived products have recently become available, such as anti-thrombin III concentrates and alpha-1-antitrypsin. These products have fairly specific indications. Antithrombin III (ATIII) is used in the treatment of patients with known hereditary antithrombin III deficiency as prophylaxis for venous thrombosis in the peripartum; it may also be used in the management of anti-thrombin III deficient patients who have an active venous thrombosis and who are not responding to heparin therapy with prolongation of the activated partial thromboplastin time (aPTT). The use of antithrombin III in acquired ATIII deficiencies, for example, in patients in the intensive care setting, consumptive coagulopathies (DIC) or prior to cardiac surgery, is not established. The dose of ATIII is approximately 0.7 U/Kg per 1% increase. Antithrombin III has a long half life and therefore can be transfused every 2-3 days unless active 'consumption' is ongoing. Alpha-1-antitrypsin is another protease inhibitor derived from human blood. It is used exclusively in the treatment of patients with severe alpha-1-antitrypsin deficiency to prevent hepatic and pulmonary disease.

31

Acute Complications of Blood Transfusion

Blood products are drugs and, as with any drugs, may be associated with adverse events. Adverse events which occur in association with the transfusion of blood products are commonly called *transfusion reactions*. Transfusion reactions are most practically divided on the basis of time of occurrence in relation to the blood transfusion. Acute complications usually occur during the transfusion event, but can occur up to several hours (4 hours) after completion of the transfusion. Delayed complications start somewhere between 24 hours and 2 weeks after the transfusion episode. Late complications may occur up to 30 years after the transfusion or series of transfusion episodes. This chapter will concern itself with acute reactions to blood transfusion.

Acute complications of blood transfusion comprise rare reactions which are potentially life threatening, and more common reactions which are nonlife threatening. These are shown in Table 32.1. For practical purposes the common nonlife threatening acute complications are seen in routine clinical practice. This is mainly because the processes involved with red cell compatibility testing in transfusion services (Chapter 7) and many of the manufacturing practices in blood centers (Chapter 2) are designed to prevent the life threatening acute complications of blood transfusion.

LIFE THREATENING ACUTE COMPLICATIONS

ACUTE HEMOLYTIC REACTIONS

The most serious adverse event associated with a red cell transfusion is the occurrence of acute hemolysis of the transfused red cells. This occurs when there is a pre-existing antibody in the recipient's plasma which reacts with the transfused red blood cells. Ordinarily, this is prevented by routine compatibility testing. Acute hemolytic transfusion reactions occur usually within five minutes of initiating the blood transfusion and primarily for this reason, early monitoring of vital signs and slowing the rate for transfusion during the first 15 minutes is common. Acute hemolytic reactions can be due to either pre-existing IgM or IgG alloantibodies. IgM alloantibodies, particularly within the ABO system, will fix complement to the terminal lytic components (C9) and give rise to intravascular hemolysis. This will cause the most severe clinical symptoms. In such hemolytic reactions, these may include fever, chills, muscle pain (backache), gastrointestinal

32

Clinical Transfusion Medicine, by Joseph D. Sweeney and Yvonne Rizk. © 1999 Landes Bioscience

Table 32.1. Acute complications of blood transfusion

I. Life Threatening—[very uncommon]

   (a) Acute hemolytic reaction (1:50,000-1:100,000)

   (b) Acute anaphylactic reaction (1:100,000-1:200,000)

   (c) Transfusion related sepsis (1:4,200-1:500,000)

   (d) Transfusion related acute lung injury (? frequency)

   (e) Acute hyperkalemia or hypocalcemia

   (f) Acute hypervolemia

II. NonLife Threatening—[common: 0.5-6%]

   (a) Febrile nonhemolytic transfusion reaction (0.5%)

   (b) Urticaria (1-2%)

symptoms such as nausea and vomiting, sometimes urticaria, shortness of breath, and hypotension. Alloantibodies of the IgG class will cause red cells to be removed extravascularly, predominantly in the spleen, and will result in less severe clinical symptoms. Other IgG alloantibodies may fix complement, but only to the third component (C3), such as antibodies within the Kidd system and, occasionally Kell system. This again results predominantly in extravascular hemolysis, but removal by liver or bone marrow macrophages may also occur. The mechanisms of acute hemolytic transfusion reactions are shown in Figure 32.1.

The pathophysiology of the clinical symptoms caused by hemolytic reactions is illustrated in Figure 32.2. Antibody and/or complement binding to red cells results in phagocytosis by macrophages and the generation of a variety of biologically active peptides, such as the inflammatory cytokines and activated complement peptides. This causes the spectrum of clinical symptoms, which will be present to a varying degree in any individual, depending on the rate and type of hemolysis. Activation of the coagulation system may cause disseminated intravascular coagulation (DIC). Renal injury arises from intrarenal shunting, causing acute renal failure; damage to lung parenchyma may cause a noncardiogenic pulmonary edema. Dysfunction in these three organs will dominate the early clinical picture in acute hemolytic transfusion reactions. It should be noted that fever may be an early manifestation of acute hemolysis.

The management of any transfusion reaction is to immediately stop the transfusion; an intravenous line should be kept open with saline. A serum specimen should be drawn and sent to the blood bank without delay, together with the administration set and the remaining untransfused red cells. Clerical checks are

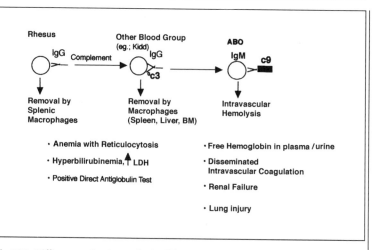

Fig. 32.1. Different mechanisms of red cell hemolysis caused by red cell alloantibodies.

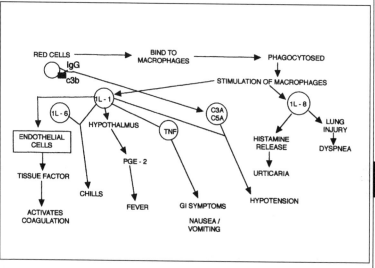

Fig. 32.2. Pathophysiology of hemolytic transfusion reactions.

32

performed to ensure that the identification of the recipient has been performed correctly. In the blood bank, visual inspection of the serum for hemoglobinemia (red-tinged serum) and performance of a direct antiglobulin (direct Coombs) test are the important tests. In severe hemolytic reactions, the serum is often red tinged, but the direct antiglobulin test may be negative since the transfused cells may all have been hemolyzed. In less severe reactions, the serum is only slightly red tinged, or not at all, but the direct antiglobulin test will be positive for either IgG, or complement, or both (Chapter 7). Thus, either of these tests will almost certainly be positive in an acute hemolytic reaction. If desired, a specimen of urine can be sent for hemoglobinuria, if there is a strong suspicion of hemolysis. It is theoretically possible that hemoglobin arising from rapidly hemolyzed red cells could be present in the urine, but absent in the serum, although, in practice, this is highly unlikely.

Active management of these patients is important. Baseline measurements should be done immediately of the prothrombin time (PT), fibrinogen, hemoglobin, platelet count, creatinine and electrolytes. Pulmonary function should be evaluated clinically. Intravenous saline (at least 100 ml/hour) with furosemide should be started to ensure a high urine output. The rationale for the use of furosemide is to reverse the intrarenal shunting, preserving the blood supply to the outer layer of the renal cortex and averting tubular necrosis. Close attention should be paid to the prothrombin time, fibrinogen level and platelet count. A fibrinogen of less than 80 mg/dL or a platelet count less than 50 x 10^9/L with evidence of clinical bleeding may require plasma and/or platelets transfusion. A renal consult should be obtained because of the potential need for dialysis and a pulmonary consult to evaluate pulmonary function.

With appropriate management above, most of the severe symptoms will resolve within 36 hours. At this time, the creatinine should be decreasing, and the fibrinogen, clotting times, and platelet count returning to normal. The platelet count will take several days, however, before returning to normal. With good clinical management, mortality from acute hemolytic reactions should be as low as 3%.

Prevention of an acute hemolytic reaction is essential, and it is for this reason that the tests involved in red cell compatibility testing are performed. In addition, proper specimen identification from the recipient at the time of sample collection and recipient identification prior to transfusion are critical (Chapters 6 and 7). A retrospective analysis of mortality association with acute hemolytic reactions shows that the most common error is failure to adequately identify the recipient, either at the time of sample collection or blood administration.

Acute anaphylactic reactions are very rare in blood transfusion. The anaphylactin which causes the reaction may be a cellular or soluble component of the transfused product. Patients exhibiting severe anaphylactic reactions should be screened for IgA deficiency. Rarely ethylene oxide, which is used to sterilize blood containers, may cause such a reaction. The management of acute anaphylactic reaction is the administration of epinephrine, antihistamines, and corticosteroids

Transfusion related sepsis is an area of current interest in transfusion practice (Chapter 35). Transfusion related sepsis in association with red cell transfusions is extremely rare and probably occurs with a frequency of approximately 1:500,000 units. Recent reports of red cell related transfusion sepsis show that autologous red cell units are far more likely (3-5 times) to be associated with this complication than allogeneic red cell units. The absolute risk, however, is still very low. Transfusion related sepsis in the case of red cells is due to blood collection from asymptomatic bacteremic donors. The most common organism is *Yersinia enterocolitica*, but other organisms such as staphylococcus have been implicated. Bacterial sepsis in association with platelet transfusion is considered to be a more common, if unrecognized, problem. Platelets are stored at room temperature (i.e., between 20-24 °C), and the higher temperature favors bacterial growth. The shelf life of platelets is limited to five days, primarily for this reason, and sepsis is rare in platelet products which have been stored for less than 3 days. Sepsis has been reported to be more common in platelets from pooled random donors than from apheresis platelets (Chapter 28); however, pooled random donor platelet products tend to be transfused later in their shelf life than apheresis platelets. Sepsis from platelet products arises mainly due to the contamination of platelets with bacteria on the skin surface at the point of venipuncture. Thus, coagulase negative staphylococci are commonly implicated. The true frequency of occurrence of such sepsis is unknown, as many patients receiving platelets are, in addition, concurrently receiving broad spectrum antibiotics, because of neutropenic fever. It is possible that these antibiotics are protective to the recipient, resulting in a mild or clinically silent reaction.

Bacterial sepsis is characteristically associated with a high fever (39°C; 103°F) and hypotension. This is sometimes helpful in separating this rare cause of fever from the far more common cause due to allogeneic leukocytes (Chapter 36). Sepsis is a potentially devastating complication of blood transfusion, and may cause death within 24 hours of the transfusion. A high index of suspicion and early energetic treatment with intravenous antibiotics is indicated. Gram stain of the blood aids diagnosis.

Transfusion related acute lung injury (TRALI) is a form of noncardiogenic pulmonary edema. TRALI has been associated with all types of blood products, but particularly with plasma or plasma containing products and dyspnea is the prominent symptom, which usually starts approximately one hour after the transfusion has been initiated. Hypotension may also be observed. The underlying cause of most cases of TRALI is the presence of (HLA) antibodies in the donor plasma reacting with antigens on the neutrophils of the recipient. These antibody coated neutrophils aggregate in the pulmonary capillaries where complement becomes activated and an inflammatory reaction ensues resulting in pulmonary edema (alveolitis). In at least 85% of recipients, the patient will recover within 48-72 hours although, in some instances, short term ventilation may be required. In a small percentage of cases (approximately 15%), more serious lung injury can occur. It is for this reason that TRALI is included among the acute life threatening complications of blood transfusion.

Acute metabolic abnormalities may occur with blood transfusion, but this is seen only in association with massive transfusion (Chapter 14). Hyperkalemia is seen only in patients massively transfused with red blood cells. This can particularly occur with red blood cells which have been irradiated since the potassium levels in these products can be very high (60-90 mEq/L) or in a recipient who has a limited ability to accommodate to a high potassium challenge, such as an infant undergoing exchange transfusion or an adult patient with renal failure. Acute hypocalcemia may occur with the transfusion of large amounts of plasma containing products, particularly fresh frozen plasma and, to a lesser extent, platelets. These products are stored in trisodium citrate and the citrate is capable of calcium chelation. Ordinarily, transfused citrate will be metabolized to carbon dioxide, giving rise to the production of bicarbonate and a metabolic alkalosis. This metabolic alkalosis tends to offset an increase in potassium by favoring a movement of potassium intracellularly. If the concentration of citrate delivered to the liver is excessive, (for example, in low weight females), acute hypocalcemia can occur with the clinical features of tetany, convulsions and hypotension. It is for this reason that surgeons and anesthesiologists sometimes prophylactically transfuse calcium gluconate. Routine transfusion of calcium gluconate to patients receiving only red cell transfusions has no basis whatsoever since most red cells are stored in an additive solution which does not contain citrate. In addition, a massive transfusion of plasma (1 unit every 5-10 minutes) is necessary before this complication is likely to occur. The administration of calcium should be restricted to the setting of massive transfusion, therefore, and calcium chloride is preferred since it readily supplies ionized calcium.

Acute hypervolemia should always be suspected in a patient with a poor cardiac status or in an elderly decompensating patient receiving a blood transfusion. Acute hypervolemia will present as acute shortness of breath and distinguishing this from TRALI may be difficult. Intravenous diuretics will improve the situation rapidly in hypervolemia, but not in TRALI. If doubt exists, hemodynamic measurements will easily make the distinction. It is primarily to prevent acute hypervolemia that blood transfusions are administered over lengthy periods of time, sometimes up to 4 hours. This is largely unnecessary for the majority of blood transfusion recipients, however.

The most important aspect of transfusion is to be vigilant for the first 15 minutes, since anaphylactic reactions, hemolytic reactions and septic reactions nearly always become evident at this time. The remainder of the unit can then be safely transfused in the majority of individuals within 1-2 hours in the case of red cells and less for plasma and platelets.

NON-LIFE THREATENING ACUTE COMPLICATIONS OF BLOOD TRANSFUSION

These complications of blood transfusion are quite common and may occur in between 0.5-6% of all products transfused; reactions to platelet transfusion may be even more common, but the severity of these "reactions" are often either clinically mild or asymptomatic. These complications result in transient morbidity in transfusion recipients, but the degree of discomfort can be considerable.

The characteristic symptoms which occur are fever, chills, nausea, vomiting, or myalgia. The mechanism by which these symptoms occur is thought to be due to inflammatory cytokines, similar to acute hemolytic reactions. Since the clinical symptoms may be similar to acute hemolysis, these reactions are called "nonhemolytic febrile transfusion reactions" (NHFTR). Although an increase in temperature of 1°C is necessary for the strict definition of a NHFTR, lesser degrees of temperature elevation or chills without fever occur commonly. The most common cause of these acute reactions is considered to be the presence of allogeneic leukocytes in the transfused products. This is largely based on data showing that the likelihood of a transfusion reaction is related to the white cell content of the blood product, and studies have shown a correlation between levels of inflammatory cytokines in platelet products, particularly interleukin-6 [IL-6] and clinical reactions in recipients of platelet transfusions. Other major inflammatory cytokines, such as IL-1, TNF, and IL-8 are also considered to be important contributors (Fig. 32.2).

Since most of these reactions are due to allogeneic leukocytes and possibly platelets, in red cell products, the use of bedside leukoreduction filters is the first line strategy in preventing these reactions. This is particularly valuable for red cell products where leukoreduction results in a dramatic decrease in the frequency of NHFTR. With platelets, it is uncertain whether bedside filtration results in a reduction in these reactions. Removal of leukocytes at the time of manufacture, called prestorage leukodepletion (Chapter 36), is more appropriate for platelet products. Reactions due to allogeneic leukocytes are generally transient (5-60 minutes) and are best treated symptomatically with acetaminophen. No other treatment is ordinarily acquired, and antihistamines have no role in management. Occasionally, severe reactions occur, particularly with platelet transfusions and more aggressive management of such patients with medication such as intravenous meperidine or steroids is used, although this is empirical. Use of prestorage filtration and, occasionally, washing of the product may be required in an attempt to prevent such reactions in some recipients (Chapter 36).

The other type of reaction which is commonly observed is urticaria. This reaction is likely due to allogeneic cell fragments or particulate matter arising from such fragments, or soluble substances present in the blood product, acting as allergens or by histamine. Interleukin-8, which can accumulate in platelet products, is also known to release histamine from basophils. Other vasoactive substances could also accumulate during storage. The management of urticaria differs from the management of a NHFTR. Patients experiencing urticaria need to have the transfusion stopped initially, in order to evaluate the clinical situation. If urticaria is the *only clinical manifestation*, there is no need to perform any further testing. The patient may be treated with an antihistamine, and the blood product recommenced slowly. It is uncertain whether leukoreduction reduces the occurrence of urticarial reactions.

One of the more common errors is the routine use of antihistamines in the prophylaxis or management of NHFTR. This is a wasteful practice which does no have a physiological basis and has no empiric justification.

Recently, acute hypotensive reactions have been reported in patients receiving bedside leukoreduction filtered blood products, particularly platelets. The mechanism has not been fully understood and all filter types have been implicated. It i best to manage these patients with prestorage leukoreduced blood products, which have not been associated with these reactions.

In all cases of a suspected transfusion reaction, the transfusion should be stopped and the clinical situation evaluated. The transfusion should never be re commenced unless urticaria is the only clinical feature, as sepsis can never be to tally excluded, even by a negative gram stain of the blood product.

Delayed and Late Complications of Blood Transfusion

In the previous chapter we have dealt with acute complications of blood transfusion which is the most common type of reaction observed in practice. Delayed reactions can be arbitrarily defined as reactions occurring between 24 hours and 2 weeks after transfusion. Late complications can arbitrarily be defined as complications, which occur from 2 weeks to 30 years after a transfusion or series of transfusions. However, there is no absolute distinction between acute and delayed reactions, and, to some extent, they can overlap, for example, patients may experience symptoms such as fever several hours after completing a red cell transfusion. These are generally considered within the spectrum of acute reactions.

DELAYED BLOOD TRANSFUSION REACTIONS

The most important delayed posttransfusion reactions are shown in Table 33.1. All these reactions are very uncommon. The most common of these reactions is the occurrence of a delayed hemolytic transfusion reaction. Delayed hemolytic transfusion reactions (DHTR) are due to the technical failure to detect an antibody present in the patient's plasma or serum prior to transfusion. This is not to imply that a procedure was performed technically incorrectly, but that all techniques have limitations in sensitivity and occasionally will result in a false-negative, i.e., failure to detect an antibody when one is present. These red cell antibodies arise as a result of transfusions or pregnancies, and the antibody increases in titer after re-exposure to the antigen on the transfused red cell. Although this may occur as early as 24 hours after the transfusion, it more commonly occurs after 2-7 days. DHTR are due, therefore, to very low levels of undetected pre-existing antibodies. The clinical features of delayed hemolytic reactions tend to be similar to, but milder than those which occur in association with acute hemolytic reactions, and rarely require hospitalization. This is because most of the antibodies are IgG and in many instances, such as the Rhesus system antibodies, do not fix complement. Intravascular hemolysis is, therefore, extremely rare. The vast majority of these reactions are silent being clinically evident only in a small proportion (20% or less). These reactions are often discovered on the basis of an unexplained hyperbilirubinemia or decrease in hemoglobin. Alternatively, these reactions are uncovered by the finding of a positive antibody screen or a positive direct antiglobulin test in the serum of a patient several days or weeks after a red cell

33

Clinical Transfusion Medicine, by Joseph D. Sweeney and Yvonne Rizk. © 1999 Landes Bioscience

Table 33.1. Delayed transfusion reactions

(24 hours-2 weeks)

(a)   Delayed hemolytic reaction

(b)   Transfusion associated graft versus host disease (Chapter 37)

(c)   Posttransfusion purpura

(d)   Transfusion transmitted protozoa (Chapter 35)

Table 33.2. Late complications of blood transfusions

(2 weeks-30 years)

(a)   Iron overload hemosiderosis

(b)   Transfusion transmitted viruses (Chapter 34)

(c)   Transfusion transmitted protozoa or helminths (Chapter 35)

(d)   Alloimmunization to red cells and HLA antigens

transfusion (in the setting where the pretransfusion antibody screen was negative). These reactions are important to investigate since they have implications for future transfusions for that particular recipient.

When clinical symptoms occur they are typically fever, chills, backache, nausea, vomiting, apprehension, etc. In rare instances, for example, with antibodies against the Kidd system, intravascular hemolysis and hemoglobinemia have been reported. DHTR have a frequency of 1:2,000 to 1:5,000. Fatal DHTR reactions are likely to have a frequency of less than 1:500,000. There is no specific treatment and only rarely is energetic clinical management required. As emphasized in Chapter 7, proper identification of blood specimens at the time of collection allows the Blood bank to trace previous records related to a patient in which the historical presence of an antibody is recorded. This will help prevent DHTR.

All other delayed reactions are exceedingly uncommon. Transfusion associated graft versus host disease (TA-GVHD) occurs between 4-20 days after transfusion. It is a devastating event and is discussed in more detail in Chapter 37. Posttransfusion purpura is another very uncommon complication of blood transfusion which occurs about 8-14 days after blood transfusion. In this situation, patients present typically with bruising or other features of thrombocytopenia, such as epistaxis. The platelet count may be extremely low, often less than 5×10^9/L.

33

The patient's pretransfusion platelet count, when available, is normal. This reaction is typically observed in cardiac surgery and, therefore, the patient presents 5-10 days after discharge. Posttransfusion purpura is due to the development of an antibody which is capable of causing premature removal of autologous (patient's own) platelets. The most common platelet antigen involved is called Pl^{A1} (HPA 1a) and patients who develop this complication lack this antigen on their platelets. The management of this disease is early recognition and treatment with intravenous gammaglobulin since a high mortality has historically been reported. Steroids may be used and plasma exchange (Chapter 40) may also be performed. The platelet count will increase to normal within 48-72 hours. Platelet transfusions *should be avoided* in these patients. Red cell transfusions may be necessary in patients who have evidence of bleeding and symptomatic anemia. Although red cells from a PL^{A1} negative donor are preferable, if at all possible, washing and filtration of the cells may be the only logistic option available, particularly in blood centers, which do have a panel of PL^{A1} negative donors.

Transfusion transmitted protozoa, causing babesiosis or malaria may sometimes produce clinical symptoms in this time period, especially in splenectomized patients. These are discussed in Chapter 35.

LATE COMPLICATIONS

Late complications of blood transfusion occur between 2 weeks to 30 years after the completion of the transfusion or series of transfusions. These complications can result from a single blood transfusion, such as a viral disease, or multiple episodes of blood transfusion.

One of the more important late complications is the development of iron overload hemosiderosis. This is a particular problem in patients with thalassemia who are transfused to maintain hemoglobins at or near the normal range (Chapter 17). Transfusion overload hemosiderosis may cause iron accumulation in many organs, but of particular concern is the development of cardiomyopathy. A single unit of blood contains 250 mg of iron and after multiple red cell transfusions, iron overload to a level well in excess of 100 g total body iron can occur. This becomes deposited in the heart, liver and other organs. Iron accumulation is best monitored by serial measurement of serum ferritin. The management of transfusion hemosiderosis is highly specialized and usually involves subcutaneous desferrioxamine, with or without vitamin C, to enhance urinary iron clearance.

Transfusion transmitted viruses (TTV) typically produce clinical symptoms during this time period. TTV are described in more detail in Chapter 34. Transfusion transmitted protozoa or helminths may also produce clinical symptoms in this time period and are discussed in Chapter 35.

Primary alloimmunization to red cell and HLA antigens occurs several weeks after transfusion with the development of antibodies of the IgM, then, subsequently IgG class. Primary alloimmunization to red cell antigens occurs in about 1% of all blood transfusions. Primary alloimmunization for practical purposes is always clinically silent, although, rarely it may be associated with laboratory evidence of hemolysis, 7-12 days after transfusion. This is most commonly seen in Rhesus (D) negative patients who are transfused with Rhesus positive (D) red cells which occurs occasionally on account of a shortage of Rhesus negative blood. Allo immunization to red cell antigens (See this Chapter, DHTR) has implications for any future red cell transfusion requirements for a patient. Alloimmunization to HLA antigens is only of practical significance if a patient requires multiple platelet transfusions (Chapter 28) or is a potential solid organ allograft recipient.

Blood Transfusion Transmitted Infections I: Viruses

The capability of blood transfusion to transmit viral disease has been known since the 1940s with the observation that plasma transfusion could cause hepatitis. The potential for blood transfusion to transmit viral diseases represents the most deep-felt fear on the part of the general public in relation to blood transfusion. While hepatitis transmission has always been the most common infection transmitted by blood transfusion, the human immunodeficiency virus (HIV) epidemic in the early 1980s and general awareness of HIV transmission by blood transfusion has caused the HIV virus to be the focal concern on the part of the general public and has impacted greatly on overall blood transfusion practice since 1982.

Viruses which are known to be transmitted by blood transfusion are shown in Table 34.1 and are arbitrarily divided into three groups for discussion purposes. The transmission of viruses which results in significant morbidity in transfusion recipients is, at present, uncommon and estimates of risk are shown in Table 34.2.

Group I includes those viruses which are present in allogeneic leukocytes only and not as free virions in plasma. These viruses, therefore, are not known to be transmitted by frozen plasma, cryoprecipitate or plasma-derived products. The most prominent viruses in this group are viruses of the Herpes family, of which cytomegalovirus (CMV) is the most important. Cytomegalovirus is discussed in detail in Chapter 38. Epstein-Barr virus (EBV; HHV-4) may be transmitted by blood transfusion but rarely is associated with significant morbidity in transfusion recipients. A more recently described virus is human herpes virus, type 6 (HHV-6). This virus is very prevalent in normal healthy donors and has been associated with severe pneumonia in bone marrow transplant recipients, but primary transmission by blood transfusion is only speculative. HHV-6 is of uncertain significance for blood recipients, as most patients in any event have been exposed to this virus in early life. An additional recently described virus is herpes virus type 8 (HHV-8). Although HHV-8 has not been shown to be transmitted by blood transfusion, it has been identified as the cofactor for Kaposi's sarcoma. In common with the other herpes viruses, however, this virus has the potential to be transmitted by blood transfusion.

Other viruses in this group are the human T-lymphotropic viruses (HTLV-types 1 and 2). These are part of the retrovirus family. Transmission of these viruses by blood transfusion is of concern because in some blood transfusion recipients they have been associated with the development of a T-cell lymphoma after an incubation period of 10-30 years, or a myelopathy after a much shorter incubation period of 2-4 years.

34

Clinical Transfusion Medicine, by Joseph D. Sweeney and Yvonne Rizk. © 1999 Landes Bioscience

---

Table 34.1. *Viruses known to be transmitted by blood transfusion*

I.   Viruses present in allogeneic leukocytes only (transmitted by red cells and platelets, *but* not transmitted by frozen plasma, cryoprecipitate or plasma derivatives).

(a) Cytomegalovirus (CMV or HHV-5)

(b) Epstein-Barr Virus (EBV or HHV-4)

(c) Human T-Lymphotrophic Virus (HTLV-1/11)

(d) Human Herpes Virus, type 6 (HHV-6)

(e) Human Herpes Virus, type 8 (HHV-8)

II. Viruses present in both allogeneic leukocytes *and* as virions in plasma (transmitted by all types of blood products).

(a) Human Immunodeficiency Virus (HIV-1; HIV-2)

III. Viruses present in *plasma* only as free virions (transmitted by all types of blood products).

(a) Hepatitis A (HAV)

(b) Hepatitis B (HBV)

(c) Hepatitis C (HCV)

(d) Hepatitis D (HDV)

(e) Hepatitis E (HEV)

(f) Hepatitis G (HGV)

(g) B19 parvo virus

Among the viruses in this group, routine tests are performed only for the HTLV viruses. The transmission of all viruses in this group by blood transfusion is likely to be either greatly reduced or eliminated by the use of leukoreduction filtration, but in practice, at this time, this approach is only used to prevent CMV infection (Chapter 38).

Group II viruses are those which are present both in allogeneic leukocytes and in plasma. They are, therefore, transmitted by all types of blood products. The most important virus in this group is the human immunodeficiency virus(es) (HIV type 1 and type 2). Since May of 1985, all blood donations have been routinely screened for the antibody to HIV-1 and, since early 1996, routinely screened for p-24 antigen, an early plasma marker of HIV infection. This has been successful in eliminating the vast majority of potentially infectious units. Regrettably, blood donors who are exposed to HIV virus may be infectious for a period prior to development of plasma markers, (either of p-24 or of anti-HIV-1). This period is sometimes described as the *serosilent window period* and constitutes the current danger for the residual small number of cases of HIV which are transmitted by blood transfusion. Further efforts in this area will soon involve the use of nucleic acid analysis of plasma to detect HIV viral RNA, and it is likely that this will further shorten the duration of the serosilent window period. It should be noted, however, that the current risk of transmission of HIV by blood transfusion is extremely low (see Table 34.2). HIV virus appears less likely to be transmitted by either washed blood components or red blood cells which are transfused later in their shelf life, i.e., stored in a refrigerator for greater than 25 days. In addition, leukoreduction of red cell products is known to cause a significant reduction in the viral load due to a reduction in allogeneic leukocytes and also platelets which contain HIV virions on their surface. These approaches are not applicable in practice, however, in attempting to prevent HIV-1 transmission.

Group III viruses are present in free plasma only and viruses in this group, like group II, may be transmitted by any type of allogeneic blood product. The most prominent viruses in this group are the hepatitis viruses. They will be discussed in alphabetical order rather that in order of clinical significance.

Hepatitis A virus (HAV) causes acute infectious hepatitis and does not have a chronic carrier state. Hepatitis A transmission by blood transfusion has only been shown to occur in association with the transfusion of plasma derivatives, such as factor VIII concentrates in patients with hemophilia. This has been related to inadequate viral attenuation steps (Chapter 31). The commonly transfused blood components such as red cells and platelets are rarely associated, if ever, with hepatitis A transmission.

Hepatitis B (HBV) is a very important virus in blood transfusion because hepatitis B has a chronic asymptomatic carrier state. In 1972, testing for hepatitis B (serum hepatitis) was the first viral test to be performed to detect hepatitis in

Table 34.2. Approximate estimates of likelihood of clinical significant viral disease transmission by blood transfusion (1999)

(Herpes Viruses Not Included)	Risk is per unit	
Any Virus	–	1:34,000
HBV	–	1:60,000
HCV	–	1:100,000
HTLV-1/11	–	1:65,000
HIV-1/2	–	1:500,000

34

blood donors. The hepatitis B test detects hepatitis B antigen. Further developments in tests for hepatitis B since that time have improved test sensitivity for detecting donors who are carriers. Hepatitis B transmission by blood transfusion results commonly in the development of clinical features of hepatitis, usually between 2-6 months after the transfusion episode. Most cases of hepatitis B will resolve spontaneously, but acute fulminant forms of hepatitis with liver failure can occasionally occur. Even when resolution occurs, approximately 10% of such patients will develop a chronic carrier state. These carriers may later develop other complications, such as cirrhosis or hepatocellular carcinoma after a period of 10-30 years. The current risk for transmission of hepatitis B is, however, quite low and estimated at 1:60,000. Although HBV remains a rare cause of transfusion transmitted hepatitis in North America and Europe, it is a far greater problem in Eastern and Southern Europe, North Africa and Asia where many first time donors are HBV positive. As in the case of HIV, plasma testing for HBV using nucleic acid amplification technology (such as PCR) will likely further reduce HBV transmission by blood transfusion.

Hepatitis C remains an important form of hepatitis transmitted by blood transfusion. In the older literature, this virus was often referred to as non-A, non-B hepatitis. The presence of hepatitis C virus is detected using an antibody test to hepatitis C virus (Anti-HCV). Since its introduction in 1990, this test has been very successful in eliminating many blood donations with the potential to transmit hepatitis C virus. A greatly improved anti-HCV test was implemented in 1992. In the late 1980s, surrogate markers for hepatitis C infection, i.e., measurements of alanine aminotransferase (ALT) and antibody to the core protein of hepatitis B (anti-HBc) were initiated in an attempt to reduce hepatitis transmission by blood transfusion and were successful in reducing some cases of hepatitis C transmission. The introduction of anti-HCV testing, however, rendered these surrogate tests less useful, although they are still used in some blood centers. Hepatitis C associated hepatitis has an incubation period of 2-6 weeks. Most cases of hepatitis C are asymptomatic (approximately 75% of cases), but a chronic carrier state develops in 50-70% of exposed patients. Chronic carriers of hepatitis C may develop a chronic hepatitis, or cirrhosis and have an increased incidence of hepatocellular carcinoma. Thus, the end result of hepatitis C infection in many cases may lead to a need for liver transplantation, 10-30 years after exposure. The current risk of hepatitis C infection per unit has decreased from a peak of perhaps 1-5% in the 1960s, to a current risk of approximately 1:100,000. Nucleic acid testing, expected to be introduced in 1999, will further reduce this risk to 1:500,000.

Hepatitis D virus (HDV or delta virus) differs from the other viruses in that it can only cause infection in recipients who are hepatitis B surface antigen positive, i.e., carriers of HBV. It is therefore only commonly seen in patients, such as hemophiliacs who have been exposed to plasma derivatives.

Hepatitis E virus (HEV) is common in Eastern Europe, Asia and in Africa. HEV has been implicated in transfusion transmitted hepatitis in underdeveloped countries, but no cases have been described in the United States. HEV is very similar to hepatitis A virus in clinical features and is not known to have a chronic carrier state.

Hepatitis G virus (HGV) is a more recently described virus with some similarity to HCV. It has a high seroprevalence rate in many blood donor populations studied (up to 4%). Infection with HGV can be associated with transient transaminasemia, but it is not known clinically to have any short or long term effects. At this time, a co-infection with hepatitis G and hepatitis C does not appear to increase the severity of hepatitis C infection. Testing for anti-HGV virus is not routine, and early information indicates that routine testing is not warranted. This situation will presumably undergo more close scrutiny as further information develops regarding this virus. Whether the liver cell is the target cell for this virus is also unsettled at this time.

The B-19 parvovirus is another virus known to be transmitted by blood transfusion. The parvoviruses are infections only for animals with the exception of B-19, which causes an acute infection, "fifth disease," in children and arthritis in adults. Nevertheless, B-19 virus transmission by transfusion ordinarily appears to rarely cause significant clinical symptoms. However, B-19 virus infection may cause a transient bone marrow suppression. This is of clinical significance in certain patients, such as patients with hemolytic states or in bone marrow transplant patients where a transient marrow failure is significant. Routine viral attenuation methods which are successful in destroying the hepatitis or HIV viruses, such as physical or chemical methods, have shown little success in inactivating the B-19 parvovirus (Chapter 31).

More recently, other viruses have been described such as TT virus (transfusion transmitted virus) but little is known regarding the significance or prevalence.

34

Blood Transfusion Transmitted Infections II: Bacteria, Protozoa, Helminths and Prions

Although viral disease transmission by blood transfusion is the dominant concern regarding the transmission of infectious diseases by blood transfusion, bacteria, protozoa, helminths and possibly other agents may also be transmitted by blood transfusion.

The most important bacteria transmitted by blood transfusion are shown in Table 35.1. Bacterial sepsis associated with red blood cells is a potentially life threatening situation (Chapter 32.1). The risk of bacteria contaminating red cell products causing a septic reaction is related to the duration of in vitro storage. More than 50% of all cases of red cell-associated bacterial sepsis are due to a single bacterial species, *Yersinia enterocolitica*. This is because *Y. enterocolitica* survives well during long periods of refrigerated storage; in addition, as red cell hemolyze during storage, the iron released is used by *Y. enterocolitica* to facilitate growth. This bacteria produces a toxin which accumulates during storage and gives rise to the clinical symptoms (fever, hypotension). Less often, species such as Pseudomonas and Salmonella have been implicated in red cell associated sepsis.

Clinical symptoms of red cell sepsis usually, but not always, occur very quickly after the transfusion is initiated. Primarily, a high fever is characteristic (>2°C; >3.5°F) and hypotension may occur; however, lesser degrees of fever or chills with or without hypotension may be the only manifestation. Unless the diagnosis is made promptly and antibiotics started immediately, a fatal outcome may occur. As the organisms are primarily gram negative, an appropriate antibiotic to administer is erythromycin, given immediately intravenously. The occurrence of fever (Chapter 32) is not uncommon in patients receiving blood transfusion, and distinguishing bacterial sepsis from other causes of fever is not possible clinically, although the extent of the fever and the presence of hypotension should always suggest bacterial sepsis.

Red cell sepsis is a rare event, occurring in approximately 1:500,000 units transfused. It should be noted that autologous blood collected preoperatively (Chapter 3), is considered to be at increased risk for this complication. Therefore, autologous blood, while safer than allogeneic blood is not entirely safe. It is also primarily because of the possibility of red cell sepsis that patients who develop fever in association with blood transfusion and are found not to be hemolyzing (Chapter 32) should not have the red cell transfusion recommenced, since sepsis cannot be excluded.

Clinical Transfusion Medicine, by Joseph D. Sweeney and Yvonne Rizk. © 1999 Landes Bioscience

Measures to prevent this unusual, but potentially fatal complication of blood transfusion are lacking, as questioning blood donors with regard to a history of recent diarrhea, for example, is largely ineffective in identifying implicated donors. Shortening the storage time of red cells from 42-25 days would be useful, but would cause difficulties with inventory management. The bacteria which contaminates red cell products are generally present in the blood at the time of collection, i.e., the donor is bacteremic but asymptomatic. As noted below, this source of bacteria is very different from the bacteria which contaminate platelet products.

Bacterial sepsis associated with platelet transfusion is considered a far more common occurrence. Unlike red cell products, the bacteria which contaminate platelet products likely originate from the skin of the donor at the time of venipuncture for blood collection. Therefore, organisms such as skin commensals are the prominent bacteria in platelet associated bacterial sepsis (Table 35.1). Platelets are also stored at higher temperatures (between 20-24°) and this facilitates the growth of many bacteria. Platelet transfusion associated sepsis occurs with platelets that have been stored for at least 3 days and, more commonly, 4 or 5 days. The clinical features of platelet associated sepsis are similar to those of red cell associated sepsis. However, many patients receiving platelet transfusions, such as leukemic patients or bone marrow transplant recipients, are concurrently receiving broad spectrum antibiotics because of neutropenic infection. Therefore, to an extent, there is protection from the transfusion of a bacterial contaminated platelet product.

The frequency of occurrence of bacterial associated sepsis with platelets is unknown since in many instances this may go undiagnosed, but it may be as high as approximately 1:4,000 transfusions. Bacterial sepsis is more commonly associated with the use of pooled random donor platelets, as discussed in Chapter 28. However, it is important to emphasize that random donor pools tend to be transfused later in storage than apheresis platelets, and this factor alone may account for this difference in product type associated sepsis.

Treponema palladium (the organism which causes syphilis) is known to be transmitted by blood transfusion and testing for syphilis using an antibody test has been performed since 1948. Transmission by blood transfusion occurs rarely at this present time. Moreover, testing for syphilis using an antibody test is ineffective in identifying infectious donors since many donors who are bacteremic at the time of donation are in the early phases of infection and seroconversion may not have occurred. Routine testing of blood serologically for syphilis, therefore, identifies individuals who have had previous infections, which are now resolved. The interest in testing blood donors for syphilis at the present time is the use of this test as a surrogate marker for HIV-1 infection and, for this reason, the test has largely continued to be used. It is interesting to note that the only bacterium for which blood donations are routinely tested does not account for any significant fraction of bacterial associated infections transmitted by blood transfusion!!

35

Lyme disease is due to another spirochete, known as *Borrelia borgdorfii*. Although this organism grows well in both red cells and, particularly platelets, no cases of Lyme disease transmitted by blood transfusion have ever been reported.

Table 35.1. Bacteria transmitted by blood transfusion

a) **Red cells:**

Yersinia enterocolitica

Pseudomonas fluoresces

Salmonella sp.

(b) **Platelets:**

Staphylococci (epidermis and aureus)

Salmonella and Serratia spp.

B. cereus

(c) **Miscellaneous:**

T. pallidum (Syphilis)

Borrelia burgdorfii (Lyme disease)

Water-bath or platelet pack contamination

On rare occasions, contamination of water-baths in which plasma has bee thawed may cause sepsis, since a break or leak in a plasma bag can allow bacteri to enter the bag. This is an unusual complication and is generally avoided in bloo banks by double wrapping of the plasma bag during thawing. Faulty manufactur ing of platelet bags has also been associated with contamination by bacteria insid the humidified environment of the blood pack, but this is rare.

Table 35.2 shows the protozoa and helminths known to be transmitted by bloo transfusion. With regard to protozoa the most important parasite is that of ma laria, particularly plasmodium malaria. The transmission of malaria by blood trans fusion is a very uncommon event in the United States, but transmission is mor common in malaria endemic zones outside of the U.S. It is a practice to question all donors with regard to recent presence in malaria endemic zones, whether the received chemoprophylaxis for malaria, or if they have had an active malarial in fection. Donors are routinely deferred for one year if they have been in an en demic zone and have had chemoprophylaxis, and for a period of three years i they have had an active, recent infection. Malaria transmitted by blood transfu sion tends to produce clinical symptoms 3-6 weeks after exposure, and classi features of malaria are present. The diagnosis is normally made by examination o the peripheral blood smear. Another important protozoa disease is Chagas dis ease, which in endemic in many parts of Central and South America. This diseas is caused by *Trypanosomiasis cruzi* (*T. cruzi*). *T. cruzi* is a concern for blood trans fusion authorities in South American countries. Some blood centers in these area

35

Table 35.2. Protozoa and helminths transmitted by blood transfusion

1. **Protozoa:**

   a) Plasmodia spp—Malaria

   b) *T. cruzi*—Chagas Disease

   c) *M. Bancroti*—Babesiosis

   d) *L. Donovani*—African Leishmaniasis (Kala-Alar)

   e) *T. Gambiense*—Trypanosomiasis

   f) *T. Gondii*—Toxoplasmosis

2. **Helminths:**

   a) *W. Bancroti*—Filariasis

---

routinely add methylene blue dye to blood products in order to kill this protozoa. In the United States, several cases of transfusion transmitted Chagas disease have been reported from blood donors who have emigrated to the United States from endemic zones, especially central America. In the Southwest United States it is not an uncommon practice to question blood donors with regard to their previous residence in endemic areas. Testing of blood using an ELISA assay for antibodies to *T. cruzi* is possible but, as yet, has not become routine. Babesiosis is a tick-borne protozoa, prominent in the Northeastern United States, particularly in the islands off Massachusetts, Rhode Island, Connecticut, and New York. This red cell intracellular parasite can be transmitted by blood transfusion but may not result in significant clinical symptoms in many recipients and, therefore, may go unrecognized. For patients who have been splenectomized, however, transfusion associated babesiosis may be a life threatening complication. In endemic zones, donors are routinely questioned with regard to a history of babesiosis; questioning with regard to recent tick bites has not been shown to be effective in preventing the transmission of this disease. African leishmaniasis, or kala-alar, is caused by a protozoa, *Leishmania donovani*. *Leishmania donovani* is known to be transmitted by blood transfusion in Africa. In the early 1990s there was concern with regard to exposure of US war personnel to a related species of leishmaniasis, known as *Leishmania tropica,* and donors who had served in this area were deferred for 18 months. *Leishmania tropica*, however, unlike *Leishmania donovani*, has never been shown to be transmitted by blood transfusion. Other protozoans such as *Trypanosomiasis gambiensi*, or African sleeping disease, have rarely been transmitted by blood transfusion. *Toxoplasmosis gondii* has been transmitted by transfusion but largely in the context of leukocyte transfusions. Studies of patients who have received multiple red cell transfusions, such as thalassemia or sickle cell disease patients, using serological testing for *Toxoplasmosis gondii* have not shown increased

35

seropositivity in multitransfused recipients compared to age and sex matched controls, indicating that the transmission of *T. gondii* by blood products is not likely to be common.

The only helminth infection well implicated to be transmitted by blood transfusion is filariasis (by the organism *Wucheria bancroti*). The microfilaria of *W. bancroti* can be seen in the peripheral blood of asymptomatic donors are capable of transmitting this infection. This is only relevant in areas where this infection is endemic; however, such as in the upper Nile areas of Egypt and Sudan.

There is considerable interest recently in the possible transmission of prion diseases by blood transfusion. Prion diseases are caused by an abnormal form of a protein termed PrP^{sen} or PrP^c, which is a normal constituent of the neurons in the central nervous system and is also expressed on the surface membrane of B lymphocytes. The abnormal form of this protein, designated PrP^{SC} or PrP^{Res}, is resistant to protease digestion. The PrP^{SC} form resembles the normal protein PrP^c, except that the PrP^{SC} protein is more unfolded. Exposure of the normal PrP^c protein to the abnormal PrP^{SC} protein causes the normal PrP^c protein to become unfolded, like the PrP^{SC} protein, and excessive accumulation of the PrP^{SC} protein then occurs with resulting cell death. Much of the attention has focused on Creutzfeld-Jakob disease (CJD) with the recent demonstration that a new variant of CJD (nvCJD) is caused by the same prion which causes a disease in cattle called bovine spongiform encephalopathy (BSE or Mad Cow Disease). The concern is that asymptomatic donors who are incubating nvCJD, could have the prion particle in blood and could transmit this disease by blood donation. A recent observation that B lymphocytes may be important in transporting this disease to the central nervous system in inoculated animals has increased interest in providing leukoreduced blood for all transfusion recipients and has contributed to the recent decision by some European countries and Canada to universally leukoreduce all cellular blood products (Chapters 36; 41).

35

Special Blood Products I: Leukoreduced and Washed Blood Products

This chapter will discuss approaches to reducing or attenuating the effects of allogeneic leukocytes or soluble substances present in cellular blood products.

All transfused blood is filtered, as each blood administration set contains an in-line filter (Chapter 7). This filter is commonly made of nylon mesh and serves the purpose of removing any large clumps of cellular debris, or clots, which formed in the blood product during storage. This nylon mesh has a pore size of 170-260μ (μ or micron is 10^{-6} meter) and therefore, will not retain single cells, small clumps of cells or particulate matter, which may arise from the degeneration of cells during in vitro storage.

Microaggregate filters (2nd generation) represented an improvement in blood filtration. Microaggregate filters were introduced in the 1970s and were either classified as depth or screen filters, depending on their mode of action. Depth filters removed particles of an average size; screen filters had a threshold (cutoff) discriminating size. These microaggregate filters were successful in removing small aggregated cell clumps, particularly the clumps of leukocytes and platelets which develop during red cell storage. The discriminating size is between 20-40 μ. Microaggregate filters will successfully remove approximately 85% of allogeneic leukocytes (present as aggregates) in stored red cells. When coupled with modifications such as the spin, cool and filter technique (whereby a unit of blood is centrifuged, then stored after centrifugation for 24 hours and subsequently transfused through a microaggregate filter), up to 95% of the leukocytes are removed. Microaggregate filters were very successful in preventing nonhemolytic febrile transfusion reactions to red blood cells.

In the late 1980s, further developments in filtration technology produced the third generation filters, and these are the most common filters in current use. In addition to a screen function, some of these filters are coated with a chemical substance resulting in the selective absorption of different cell types. They are therefore, capable of removing large numbers of single cells, in addition to small cell clumps. These filters consist of polyester or polyurethane layers contained in a polycarbonate housing. The polyester fibers are coated with proprietary chemical material, which confer the selectivety in cell absorption. The red cell leukoreduction filters will remove both leukocytes and platelets. Platelets may actually facilitate the removal of certain types of leukocytes, particularly granulocytes. The platelet leukoreduction filters selectively remove only white cells (and not platelets!). Third generation filters are successful in removing between 99.5-99.9% of allogeneic leukocytes and therefore, the residual leukocyte load transfused is often extremely

Clinical Transfusion Medicine, by Joseph D. Sweeney and Yvonne Rizk. © 1999 Landes Bioscience

low. Most of the currently available clinical data from leukoreduction studies have used third generation filters.

More recent technological developments have resulted in the production of fourth generation filters. Fourth generation filters function like third generation filters but are capable of a greater reduction in leukocytes (higher capacitance). It is not uncommon for these filters to remove 99.99% of leukocytes and filters are currently available, which may remove 99.9999%. Although the percent removal is important, leukoreduction is generally expressed as a residual white cell content. Blood products which contain less than 5×10^6 residual white cells (1×10^6 in Europe) are called "leukoreduced" or "leukodepleted", since below this threshold febrile reactions and the risk for HLA alloimmunization or CMV transmission is greatly reduced. The degree of leukoreduction necessary to achieve these beneficial effects is shown in Figure 36.1.

Whole blood filters, which filter the whole blood donation, have recently been introduced. Most of these filters are modifications of the red cell filter and, therefore, remove platelets in addition to leukocytes. Only two components, a red cell concentrate and plasma, can be manufactured. One filter shows promise in its ability to selectively retain leukocytes, thus allowing the platelets to pass through and, therefore, all three components, red cells, platelets and plasma can be produced.

Filtration is not the only technology which is capable of leukoreduction. Modifications to apheresis devices which produce single donor platelets (Chapter 28) may also result in platelet products with low levels of residual white cells.

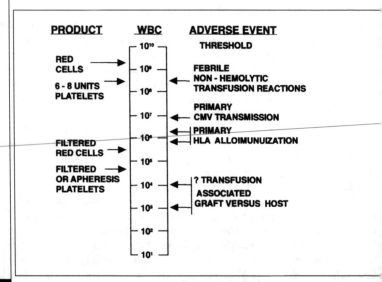

Fig. 36.1. Approximate content of allogeneic white blood cells (WBC) in different blood products and the threshold content below which the adverse event is greatly reduced. Note the scale is logarithmic.

Table 36.1. Blood filters and degree of leukoreduction

Generation of Blood Filter	Degree of Leukoreduction	Purpose
I. Standard, Nylon Mesh (170-235 μ)	none	Removal of large clumps, clots
II. Microaggregate Filters (20-40 μ)	80-95%	Removal of smaller cell clumps
III. "Third Generation" (20 μ; surface absorption)	99.5-99.9%	Selective removal of individual cell types
IV. "Fourth Generation" (high capacity; improved absorption)	99.99-99.9999%	Improved capacity over III.

Lower threshold range of particles removed are in parenthesis. μ = micron or 10^{-6} meter

An area of active interest with regard to blood filtration is the question of whether blood should be filtered at the time of manufacture (which is called prestorage leukoreduction), or filtered at the bedside (which is called poststorage leukoreduction). Prestorage leukoreduction is more likely to achieve the best reduction in transfusion reactions since there is limited potential for white cells or platelets to degenerate or form particulate matter during storage. In addition, the red cell loss is less than occurs with poststorage leukoreduction. Prestorage leukoreduction is also more convenient for the transfusionists since difficulty with blood flow sometimes occurs when bedside filtration is used. It is likely, therefore, that prestorage leukoreduction will be the dominant product in the future for patients requiring leukoreduced products. Approximately 15% of all red cells transfused in the United States are transfused as leukoreduced cells; this figure is closer to 40-60% for platelet products. The use of a third or fourth generation blood filtration adds an additional cost to the blood of USD 15-30 depending on the type of filter used. It is likely, however, that this cost will decrease with time, as leukoreduction becomes more common. Blood filtration, in addition to adding costs, may have other disadvantages. Filtration of a red cell product, for example, will remove between 4-15% of red cells and thus, the effective dose of the transfused product is reduced. Platelet leukoreduction filters may remove between 6-20% of the platelets, thus similarly reducing potency (dose). In addition, there has been some recent concern that certain bedside filters have been associated with acute hypotensive reactions in transfusion recipients. Use of prestorage leukoreduced products is preferable for such patients.

Table 36.2 shows the patients likely to benefit from the use of leukoreduced products. Most of the clinical situations have been described in other chapters and will not be described in detail here. Of recent interest is the decision by several European countries and Canada to mandate universal leukoreduction of all blood products. This is partly as a precautionary measure to prevent prion disease transmission by blood transfusion (Chapter 35) and may influence the U.S. to take a similar approach in 1999.

36

Table 36.2. Indications for leukoreduction

1. To prevent acute febrile, nonhemolytic reactions (Chapter 32).

2. To prevent primary alloimmunization to HLA antigens (Chapter 28;33).

3. To prevent primary CMV disease (Chapter 37).

4. To prevent/reduce undesired immunomodulation caused by blood transfusion (Chapter 13).

5. If a sufficient degree of leukoreduction is achieved, to prevent graft versus host disease (Chapter 36).

6. Possibly, to prevent prion disease transmission by blood transfusion.

Table 36.3. Indications for washed blood products

1. Patients with repeated acute nonhemolytic febrile or urticarial transfusion reactions, despite the use of prestorage leukoreduced blood products.

2. Patients with IgA deficiency, who have severe reactions to blood products, in the absence of blood components from IgA deficient donors.

3. Red blood cell transfusion in posttransfusion purpura, in the absence of Pl^{A1} negative blood.

4. In pediatrics, to prime extracorporeal circuits, such as cardiopulmonary bypass or therapeutic apheresis or in exchange transfusions.

Washing of blood products is becoming increasing uncommon. In the past, washing was often performed to remove allogeneic leukocytes, but this is a very inefficient way of removing white cells, achieving only a 70-85% reduction at best. Filtration (see above) is a far more effective technology.

The selective groups of patients for whom washed blood products may be useful are shown in Table 36.3. Patients who exhibit repeated nonhemolytic febrile transfusion reactions or urticarial reactions despite the use of prestorage leukoreduced blood products may be candidates for these products in order to prevent such reactions. Patients with IgA deficiency who exhibit allergic type reactions are best managed with the use of blood components from IgA deficient donors. In the absence of such products however, washed products could be transfused, although awareness of the need to intervene immediately to treat an acute anaphylactic reaction is required. In posttransfusion purpura (Chapter 33), use of Pl^{A1} negative blood components is desirable. In the absence of such components, the use of washed red cells may be appropriate. There are specific uses in pediatrics, in order to remove glucose or K^+ from the supernatant of the stored red cells, but this is only relevant in the context of massive neonatal transfusion.

It is clear that there are very limited indications for washed blood products. Requests for washed blood products often indicate a lack of awareness of physicians of the availability of other more effective products in the management of specific patient problems.

36

Special Blood Products II: Irradiated Blood Products and Transfusion Associated Graft Versus Host Disease

The irradiation of blood products using high-energy radiation (gamma rays) is exclusively performed to prevent a rare but fatal complication of blood transfusion, known as transfusion associated graft versus host disease (TA-GVHD). Other forms of radiation involving the ultraviolet region (UV) have been used experimentally alone or in combination with photochemical agents to inactive leukocytes or microbes in blood products, but are not in routine use at this time. These blood products could also be called "irradiated blood". It is important to appreciate that only blood products, which contain viable cells, need irradiation, and it is a common error to request irradiation of acellular products such as plasma or cryoprecipitate.

TV-GVHD occurs because the transfused allogeneic leukocytes (an unintended part of the transfusion component) may be viable at the time of transfusion and may undergo multiplication. Under normal circumstances, when red cells or platelets are transfused, the "passenger" allogeneic leukocytes are capable of being detected in the circulation for several hours. These allogeneic leukocytes attempt to divide as they react immunologically to foreign antigens in the host, and multiplication of these cells can sometimes be observed 3-5 days after the transfusion using sensitive techniques, such as amplication of HLA-DR genotypes using the polymerase chain reaction (PCR). Under normal circumstances, however, the host (recipient) immunocytes are successful in eliminating the donor allogeneic leukocytes and thus no observed adverse clinical event is evident. In patients with a compromised immune system, however, the ability of the host immune system to destroy the donor allogeneic leukocytes is impaired, giving rise to a situation in which the donor immunocytes proliferate and recognize host tissue as foreign. A similar situation can arise uncommonly in a nonimmunocompromised recipient, if there is similarity between the HLA type of donor and the recipient. In this situation, the donor cells are not recognized as "foreign" by the host (recipient) immune system. For practical purposes, this type of situation (known as a one way HLA match) is almost always encountered when the donor is homozygous for a HLA haplotype, which is also present in the recipient.

Regardless of the mechanism by which this rare event occurs, TA-GVHD exhibits clinical manifestations 4-21 days after the transfusion and is, therefore, a delayed adverse reaction to blood transfusion (Chapter 33). TA-GVHD affects the liver, skin, and gastrointestinal tract, giving rise to hepatitis, skin rashes, erythema and diarrhea. These clinical manifestations are similar to a graft versus host reaction

37

occurring in allogeneic bone marrow transplantation. The distinguishing feature of TA-GVHD, however, is the presence of pancytopenia due to bone marrow failure. Pancytopenia is the hallmark of TA-GVHD, accounting for the high mortality rate of at least 90%. The only potentially effective treatment is an emergency allogeneic bone marrow transplantation, which, under these circumstances, is rarely successful. It is on account of this high mortality that prevention is essential. Prevention is achieved by irradiating cellular blood products with gamma photons to a dose which renders allogeneic lymphocytes incapable of mitotic division. This can be achieved at doses as low as 500-600 rads but conventionally, the minimum dose recommended is 2,500 rads to the midplane of the blood product.

Irradiation of red cell products is known to damage the red cell membrane giving rise to an increase in extracellular potassium and a slight reduction in the recovery of red cells (a loss of about 7-8%). Irradiated red blood cells should not be stored for longer than 28 days after irradiation, or the end of the expiration period, whichever comes sooner. The potassium level at this time can sometimes be as high as 100 mEq/L, but is usually about 50-70 mEq/L. In practice, however, as discussed previously, this high potassium is only a problem in the context of neonatal exchange transfusion, massive transfusion in trauma or patients with impaired renal function (Chapter 14). It is best to irradiate red cells immediately before transfusion, if possible, as this minimizes the irradiation induced changes. Gamma irradiation at these doses has only a minimal effect on platelets and changes in potassium do not occur. Nevertheless, it is advisable to irradiate platelets immediately before transfusion, if practical, as with red blood cells.

Table 37.1 shows the major indications for the irradiation of cellular blood products. It is important to distinguish two clinical situations. First, blood is irradiated for some patients who are immunocompromised. These are patients with hereditary T-cell deficiencies; fetuses and premature infants; neonates, particularly low birth weight newborns, and some patients with pediatric malignancies such as young children with neuroblastomas. Patients with Hodgkin's disease requiring blood transfusion should routinely receive irradiated blood on account of the known T-cell immune defect—regardless of whether these patients exhibit any clinical evidence of a cell mediated immune defect. Bone marrow transplant recipients should receive irradiated blood products under certain circumstances. Candidates for allogeneic transplants should receive irradiated blood just prior to the start of the conditioning regimen until at least two years postsuccessful engraftment, and it is not uncommon to routinely transfuse irradiated blood products to these patients indefinitely after transplantation. Candidates for autologous transplants should receive irradiated blood products: (1) two weeks prior to any stem cell collection either by apheresis or bone marrow harvest. The rationale for this recommendation is that allogeneic leukocytes present from a recent transfusion could be harvested during the stem cell collection, subsequently cryopreserved, and re-infused with the transplant. (2) Autologous stem cell transplant patients should receive irradiated blood from the start of the conditioning regimen until after engraftment. After successful engraftment of an autologous transplant, it is questionable whether irradiated blood is required. However, it is not an uncom-

Table 37.1. Irradiated blood products

(I)   Irradiation for recipient reasons, i.e., immunocompromised patients.

   (a)   Hereditary Immune T-Cell Deficiencies

   (b)   Fetuses: and Neonates

   (c)   Pediatric Malignancies (e.g., neuroblastomas)

   (d)   Hodgkin's Disease

   (e)   Bone Marrow Transplants:

      1.   *Allogeneic:* from the start of conditioning regimen until *at least* two years post successful engraftment.

      2.   *Autologous:* Two weeks prior to any stem cell collection by apheresis/bone marrow harvest *and* from the start of conditioning regimen until three to six months after engraftment.

(II) Irradiation for product reasons:

   (a)   Directed donations, where the donor is *genetically related* to the recipient

   (b)   HLA matched platelets

---

mon practice to routinely administer irradiated blood for a period of 3-6 months after engraftment. Thereafter, the immune system should be reconstituted and capable of preventing TA-GVHD.

The second clinical situation is irradiation of blood products because of product type. Directed donations from donors who are *genetically related* to the recipient should all be irradiated prior to transfusion. All HLA matched platelets should be irradiated. In many cases, however, HLA matched platelets are irradiated in any event since the recipient may be immunocompromised.

Table 37.2 shows clinical situations where blood is sometimes routinely irradiated, but where the data demonstrating the appropriateness of this practice is lacking. First, patients with AIDS, although clearly having an immunocompromised state, do not need to routinely receive irradiated blood; (this is discussed in more detail in Chapter 12). In some clinical centers, all adult patients with hematological malignancies are routinely given irradiated blood products and in some pediatric oncology units it is common practice to irradiate blood products for all recipients with hematologic and nonhematologic cancers. Although there are some accepted indications in pediatric oncology for the use of irradiated products, the routine use of irradiated products in the treatment of the common liquid tumors, such as leukemias and lymphomas, and for many of the solid tumors (such as Wilms disease), would appear to be unjustified. In the past, granulocyte transfusions have been implicated in causing TA-GVHD in patients with acute leukemia,

37

Table 37.2. Clinical situations where blood is sometimes requested as irradiated, but where inadequate data exists to justify the practice as routine

(a) AIDS Patients (Chapter 20)

(b) Solid Organ Allografts Recipients (Chapter 12)

(c) All Adult Hematologic Malignancies (Chapter 15)

(d) All Pediatric Malignancies (hematologic and solid)

(e) All Granulocyte Products

and for this reason is has become common practice to routinely irradiate granulocytes. However, granulocytes should not be irradiated simply because of the large dose of viable immunogeneic leukocytes. Granulocytes should be irradiated if there is a *recipient indication* for the irradiation. In practice, however, granulocytes are used for severely septic and neutropenic patients, and it is often difficult to exclude a possible immunocompromised state.

Cellular blood products have been implicated in the causation of TA-GVHD only when the product had been stored for less than 14 days. Although all platelets transfused are stored for five days or less, many red cell products are transfused beyond this storage period. Theoretically requesting older red cells could be adequate prophylaxis for TA-GVHD, but in practice, this is cumbersome, has potential for error and should be avoided.

Special Blood Products III: Cytomegalovirus Low Risk Blood Products and the Prevention of Primary CMV Disease

Serological evidence of previous cytomegalovirus (CMV) infection is present in a very significant number of blood donors. Depending on geographic location and donor age, this varies between 25-70%. Thus, the potential for blood products to transmit CMV infection is high. Only subpopulations of blood donors who are serologically positive for CMV are capable of transmitting CMV, however, perhaps as few as 5%. Regardless, this subpopulation of CMV seropositive donors cannot be accurately identified prospectively at this time. As discussed in Chapter 34, CMV belongs to the Herpes group of viruses and the reservoir for this virus in asymptomatic donors is circulatory T lymphocytess. The ability to transmit CMV by blood transfusion, however, may be correlated with the presence of the virus in either donor monocytes or granulocytes at the time of donation. It follows, therefore, that leukoreduction has the potential to eliminate this complication.

CMV infection must be distinguished from CMV disease. The transmission of CMV by blood transfusion (CMV infection) is common, largely asymptomatic, and can only be recognized by subsequent serological testing. CMV disease, however, is a potentially catastrophic clinical complication. Severe pneumonia, gastroenteritis or retinitis characterizes CMV disease and can result in a fatal outcome in certain transfusion recipients, for example, patients undergoing allogeneic transplantation. Prevention, therefore, is critical. Primary CMV disease refers to the occurrence of CMV disease in a seronegative recipient; secondary CMV disease represents reactivation of CMV in a CMV seropositive recipient. CMV low risk products are used to prevent primary CMV disease and have no known role in preventing secondary CMV disease.

There are several different approaches to the prevention of primary CMV transmission by blood transfusion, as shown in Table 38.1. Historically, the most widely accepted practice is the transfusion of blood products from donors known to be serologically negative for CMV at the time of donation. Leukoreduction however, using third or fourth generation filters (Chapter 36), has also been shown to be effective in preventing CMV infection. The use of frozen deglycerolized red cells in renal transplant patients is known to be effective, but, in practice, frozen red cells are rarely used to prevent CMV transmission and, thus, the choice of product

38

Table 38.1. Types of CMV low risk blood products

---

a) Donations which are serologically negative for CMV

b) Leukoreduction by a third or fourth generation filter (Chapter 36)

c) Frozen - deglycerolized red blood cells (Chapter 39)

d) Blood products which do not contain viable white cells
    e.g. frozen plasma or cryoprecipitate

Table 38.2. Indications for CMV risk reduced blood products

(a) Intrauterine transfusions and low birth weight (<1.2 kg) infants

(b) Pregnant females of unknown or CMV seronegative status

(c) Bone marrow transplantation:

     (I)   Allogeneic transplantation of CMV negative stem cell product to a CMV
           negative recipient (both potential candidates and identified recipients).

     (II) Autologous transplantation in a CMV negative patient

(d) Solid Organ Transplantation:

     CMV Seronegative candidates

(e) T cell-immunodeficiency state in a CMV seronegative patient e.g. Acquired Immune
     Deficiency Syndrome; Severe Combined Immunodeficiency Disease.

---

generally is between CMV seronegative components and leukoreduced blood. Studies in the last ten years have shown leukoreduced blood to be essentially equivalent to CMV seronegative blood components. CMV transmission has similarities with irradiated blood (Chapter 37) in that only cellular blood products transmit CMV and red cells stored for more than 14 days are not known to transmit CMV. Frozen plasma or cryoprecipitate are CMV low risk products and, therefore, do not need to be either filtered or manufactured from donations, which are CMV seronegative.

The appropriate patient populations who should receive CMV low risk blood products are shown in Table 38.2. All intrauterine transfusions and transfusions to low birth weight infants (< 1.2 Kg) should be CMV low risk; either a seronegative donor or a leukoreduced product is acceptable for this purpose. Pregnant females of unknown CMV status or who are known to be CMV seronegative should receive CMV low risk products. The occurrence of primary CMV infection in pregnancy can have catastrophic complications for the fetus, especially in the first trimester. One of the major indications for CMV *low risk* products is in the management of patients undergoing bone marrow transplantation. Potential allogeneic transplant patients who are CMV seronegative and who will receive a CMV seronegative stem cell (bone marrow) product should receive CMV low risk products. In this patient population, there is a strong preference for the use of blood products from serologically negative donors, although current data would indicate that a leukoreduced product is equivalent. Autologous transplants need only receive a CMV low risk product if the patient-donor is known to be CMV seronegative. A leukoreduced product is generally acceptable. CMV disease is less commonly observed in autologous transplantation. Solid organ transplants constitute another important population of patients for whom CMV low risk products (Chapter 12) are appropriate if the recipient is CMV seronegative. CMV low risk products are not known to be useful in CMV seropositive allograft recipients. Although CMV disease may occur due to a different strain of CMV ("second strain CMV"), this second strain of CMV virus has only been shown to be of allograft origin and not transfused derived. A last miscellaneous group constitutes patients with T-cell immunodeficiency status, such as patients with HIV or severe combined immunodeficiency disease. Although most patients with HIV infection (approximately 85-90%) are CMV seropositive at the time of diagnosis, there is still a subpopulation who are CMV seronegative, and the occurrence of primary CMV transmission in this population needs to be avoided. Leukoreduction is a reasonable approach in these patients.

Special Blood Product IV: Frozen Blood

Frozen blood most commonly refers to red cell products which has been cryopreserved (products such as plasma and cryoprecipitate are routinely frozen). Red blood cells are cryopreserved using either 20% or 40% glycerol as the cryoprotectant; 40% glycerol is more widely used because of an extended shelf life of up to 10 years. The storage temperature is –95°C or less. Platelets can also be cryopreserved, but a different cryoprotectant, diethyl sulfoxide (DMSO) is more widely used. Platelet cryopreservation is, however, not common and has only found application in patients with acute leukemia in remission, who may subsequently undergo either consolidation therapy or bone marrow transplantation, particularly patients who have become alloimmunized to platelets. Cryopreservation of autologous stem cells (either of peripheral blood or bone marrow origin) is routine. However, stem cell collection, processing, storage, and transfusion are not within the scope of this handbook.

For practical purposes, therefore, this chapter will discuss only frozen red blood cells, often called "frozen blood". The indications for cryopreservation of red blood cells are shown in Table 39.1. Blood from a donor with a very Rare phenotype (rare blood group), when the frequency is less that 1:500 is an important indication. Such a blood type is usually in short supply and demand is variable and unpredictable. The second clinical situation is more common and arises when the donor(s) is known to lack red cell antigens to which alloimmunization is common. For example, patients with sickle cell disease (Chapter 17) have a tendency to form multiple alloantibodies after red cell transfusions. Red cells from patients with sickle cell disease frequently lack common Rhesus antigens, designated C and E, and other minor antigens such a Kell (K), Kidd (JK^b), and both of the two common antigens within the Duffy system (Fy^a, Fy^b). It is difficult in practice for transfusion services and often blood centers to have this blood available on demand in the liquid state. The third indication occasionally used for cryopreservation of blood is autologous blood donation. Two different clinical situations can occur here: the autologous blood donor with a rare blood group or multiple antibodies, where cryopreservation may be appropriate. A second situation is where surgery is unexpectedly canceled. Although in practice, blood centers do not like to cryopreserve autologous blood because of cost and logistics, there may be exceptions; for example, when elective surgery has to be deferred for approximately 3-4 weeks, or where two planned procedures separated by only a few weeks are anticipated. Without cryopreservation, the patient could undergo surgery in an anemic state with allogeneic transfusion possibly required. If the surgery can be deferred for a more extended time period, e.g., three months, it may be better to discard the

Clinical Transfusion Medicine, by Joseph D. Sweeney and Yvonne Rizk. © 1999 Landes Bioscience

---

Table 39.1. Possible indications for cryopreservation of red cells (frozen blood) and platelets

---

I. Red Blood Cells

    (a) Rare phenotype blood (rare blood group < 1:500)

    (b) Red cells known to lack multiple antigens, to which alloimmunization is common

    (c) Autologous blood prior to an elective procedure

    (d) CMV risk reduction (Chapter 38)

II. Platelets

    (a) Autologous bone marrow donor; prior to transplantation

    (b) Acute leukemia in remission, prior to consolidation therapy *or* allogeneic bone marrow transplantation

---

39

liquid blood without cryopreservation and recommence a predeposit schedule prior to the new intended date of surgery (Chapter 3). Cryopreservation of autologous blood with the intention for use in an emergency situation, such as trauma, is costly, inappropriate and logistically impractical. The need to transfuse blood urgently will not allow the time consuming task of deglycerolization and hence, this practice should be discouraged.

As indicated in Chapter 38, frozen deglycerolized red cells are known to be associated with a reduction in the likelihood of transmitting CMV disease. In practice, however, this is not a common indication for the use of this blood product.

Frozen blood takes longer to prepare for transfusion than liquid stored blood. This is because the process of deglycerolization and washing requires 45-60 minutes at a minimum, and the blood is frequently stored at a site distant from the intended site of transfusion, resulting in a transportation delay. In addition, after deglycerolization, the expiry period is reduced to 24 hours. Therefore, clinicians requesting this product should be reasonably sure of the likelihood of a transfusion in order to avoid wasting a scarce resource.

Special Blood Products V: Therapeutic Phlebotomy, Apheresis and Photopheresis

Therapeutic phlebotomy, therapeutic apheresis and photopheresis all have in common the withdrawal (removal) of blood for therapeutic purposes. The blood is either discarded (therapeutic phlebotomy); certain components are removed and discarded; (therapeutic apheresis), or are subjected to processing ex vivo before being returned to the patient (photopheresis). Thus, all these procedures are similar in principle.

THERAPEUTIC PHLEBOTOMY

The indications for therapeutic phlebotomy are shown in Table 40.1. The most common of these conditions is idiopathic hemochromatosis. Periodic removal of red cells is essential in this condition in order to prevent iron damage to the parenchyma of the liver, heart, pancreas, and endocrine organs. A program of weekly to twice weekly, phlebotomy is commenced, as tolerated by the patient, until a prescribed amount of iron has been removed (1 unit = 250 mg iron). The blood is removed in single whole blood units, although two-unit removal may be well tolerated based on experience with healthy donors. Blood collected from patients with hemochromatosis generally is discarded but could be used in theory as a red cell product for a transfusion recipient. There are several reasons not to use this product as an allogeneic red cell. First, patients with hemochromatosis may have subclinical liver damage and, thus, have elevated alanine aminotransferase (ALT) levels. In some blood centers, this will result in the blood being discarded in any event. Second, the donor's motivation for phlebotomy is personal gain and not altruism, and the donor history may be unreliable. Third, the blood in hemochromatosis is iron rich, and this is an environment in which *Yersinia enterocolitica* (Chapter 35) will grow avidly. Fourth, since this blood would need to be labeled, as from a patient with hemochromatosis; many physicians would be reluctant to prescribe this product.

Polycythemia constitutes the second important indication for therapeutic phlebotomy. These patients are phlebotomized aggressively, until the hematocrit falls below 45. Other treatments, such as chemotherapy may be used concurrently to decrease white cells, platelet counts, or both. Last, porphyria cutanea tarda (PCT) is a rare form of hereditary prophyria which is also managed by therapeutic phlebotomy. In PCT, there is often accumulation of iron in the liver, giving rise to liver

---

Table 40.1. Indications for therapeutic phlebotomy

(a) Hemochromatosis

(b) Polycythemia Vera

(c) Porphyria Cutanea Tarda

---

hemosiderosis. Phlebotomy in PCT also removes a source of the uroporphyrin, which accumulates in red cells in this condition.

40

THERAPEUTIC APHERESIS

Unlike therapeutic phlebotomy in which a whole unit of unanticoagulated blood is removed and discarded, therapeutic apheresis draws anticoagulated blood, which is then subjected to processing ex vivo with the selective return of components. Therapeutic apheresis has evolved over the last 30 years and there are now clearly defined indications for this procedure, as shown in Table 40.2. The most important accepted indication is in the treatment of hyperviscosity syndrome, either due to an increase in soluble plasma proteins; such as in myeloma or Waldenströrm's macroglobulinemia, or hyperleukocytosis with a high blast count (in excess of $50 \times 10^9/L$). For patients with myeloma or Waldenström's macroglobulinemia, there should be clinical evidence of hyperviscosity such as retinal changes, pulmonary or cerebral symptoms and the serum viscosity elevated, at least 3 or greater. For patients with acute leukemia and hyperleukoctyosis, clinical features of either pulmonary or cerebral dysfunction should be present and the blast count should be in excess of $50 \times 10^9/L$. High white cell counts in acute leukemia are often very well tolerated by these patients, and therefore, hyperleukocytosis in itself does not constitute an indication for therapeutic leukopheresis. The next important indication for therapeutic apheresis is in the treatment of thrombotic thrombocytopenic purpura (TTP) or the hemolytic uremic syndrome (HUS). As in the hyperviscosity states, urgent therapeutic apheresis is required in TTP and HUS. The availability of therapeutic apheresis has revolutionized the management of patients with this rare disorder and a rapid improvement in clinical symptoms can occur within 48 hours after initiating treatment, although usually a longer period of treatment is required. The response is less predictable in patients with HUS, and more prolonged treatment is often necessary. Patients with TTP usually require multiple treatments on consecutive days, until a response occurs, then are subsequently apheresed on alternate days or twice weekly for up to 4-6 weeks. Patients with hemolytic uremic syndrome often require more extended treatments and in many instances, complete remission does not occur. The volume of plasma exchanged daily in these conditions is generally 1.5 plasma volumes per treatment episode. Remission is best monitored by the return of the platelet count to normal and improvement in clinical symptoms.

40

Table 40.2. Indications for therapeutic apheresis

I. Accepted

a) Hyperviscosity syndrome

   1) Multiple myeloma or Waldenström's macroglobulinemia (plasma exchange)

   2) Acute leukemia with a high blast count (>50 x 10^9/L) and clinical evidence of leukostasis (leukopheresis)

b) Thrombotic thrombocytopenic purpura (TTP) or hemolytic uremic syndrome (HUS) (plasma exchange)

c) Antiglomerular basement membrane disease

d) Myasthenia gravis (plasma exchange)

e) Acute Guillain-Barre syndrome (plasma exchange)

f) Sickle cell anemia (red cell exchange)

   1) Acute chest syndrome

   2) Cerebrovascular events

   3) Priapism

g) Staph Protein A immunoabsorption

   1) Refractory immune thrombocytopenia

   2) Drug inducted HUS

   3) TTP unresponsive to plasma exchange

II. Possible:

a) Immune-complex disease with vasculitis

b) Solid organ rejection after cardiac or liver allografting

c) LDL-apheresis for elevated LDL-cholesterol, refractory to diet and cholesterol lowering drugs (LDL–cholesterol > 200 mg/dl)

d) Platelet pheresis in thrombocytosis

Acute antiglomerular basement membrane (anti-GBM) disease constitutes another specific indication for plasma exchange and should be performed in patients before the need for dialysis. The plasma exchange in anti-GBM disease is beneficial for the renal component, and it is not known to be effective for the pulmonary component of this disease. Plasma exchange in this condition is performed, therefore, to avoid or avert dialysis. Myasthenia gravis and acute Guillain-Barré syndrome are neurological conditions for which apheresis is known to be effective in causing clinical remission and shortening hospitalization. Myasthenia gravis patients may be treated if they are unresponsive to conventional medication or are being prepared for surgery, such as thymectomy. Acute Guillain-Barré syndrome patients should be treated as soon as possible after diagnosis. There are various apheresis regimens used in the management of patients with Guillain-Barré syndrome; either daily plasma exchange for four days, followed by alternate day treatments until clinical resolution or stabilization. Some regimens use alternate day plasma exchanges for a total of only two or three plasma exchanges. The optimal approach is unclear at this time. Recent information indicates that intravenous gammaglobulin (IVGG) is useful in the treatment of both Guillain-Barré syndrome and myasthenia gravis. Comparisons of IVGG with plasma exchange have shown essentially equivalent results. Thus, either form of treatment is acceptable in the treatment of these conditions. Patients unresponsive to IVGG may be candidates for therapeutic apheresis.

Another important use of apheresis is red cell exchange. In red cell exchange, autologous red cells are removed and exchanged with allogeneic red cells. The most widely used application is the treatment of sickle cell anemia in patients with the acute chest syndrome (Chapter 17). Rapid resolution may occur in this life-threatening situation. Sickle cell patients presenting with cerebrovascular events or refractory priapism may also be appropriate candidates for red cell exchange.

A variation of therapeutic plasma exchange is the use of a staphylococcal protein A immunoabsorption column. In this treatment, a volume of plasma, (between 500-1,500 ml), is separated ex vivo by centrifugation. It then percolates over a silicon column to which staph protein A has been attached. Staph protein A absorbs all classes of IgG (except IgG3), but more importantly, removes IgG containing immune complexes from plasma. This therapy has been shown to be effective in patients with refractory immune thrombocytopenic purpura (ITP) in the treatment of drug-induced hemolytic uremic syndrome, in bone marrow thrombotic microangiopathy, and in some cases of TTP, apparently unresponsive to plasma exchange and, more recently in rheumatoid arthritis.

Recently, a variation of the apheresis technique called LDL (low density lipoprotein) apheresis has become available. In this technique, plasma is percolated through dextran sulphate columns and the LDL removed. This is a very expensive form of treatment and only suitable for patients with documented persistent elevated hypercholesterolemia (LDL cholesterol > 200 mg/dl) despite the use of appropriate maximum dose of lipid lowering medications and strict adherence to diet. Under these circumstances, some patients show benefit from twice monthly LDL apheresis with regression of atherosclerotic vascular disease. Other medical conditions continue to be treated with plasma exchange, such as immune complex diseases with vasculitis. It is unclear whether these patients benefit from the plasma exchange therapy. Solid organ allograft rejection, for example in cardiac or liver transplantation, has also been managed by plasma exchange, but only anecdotes indicate clinical benefit.

Therapeutic platelet pheresis is a procedure in which there is selective removal of platelet rich plasma. This is sometimes requested in patients with high platelet counts (> 1000 x 10^9/L). In most instances, platelet pheresis is not indicated as a relationship between the platelet count and clinical thrombosis is not present. A subpopulation, however, of these patients who present with digital Ischemia or cerebrovascular events may be appropriate candidates for urgent platelet pheresis.

40

PHOTOPHERESIS

Photopheresis is another variation of apheresis in which the white cell component is exposed to ultraviolet radiation ex vivo. In this technique, a photoactive dye such as psoralen (8-methoxypsoralen or 8 MOP) is taken by mouth. Several hours later, the apheresis procedure is performed. Ex vivo, the white cell component is separated and exposed to ultraviolet radiation causing drug activation. Although the precise mechanism of action is not understood, it is considered that the photochemical reaction causes both a cell membrane and a nucleic acid effect. The only clearly accepted indication for photopheresis is in the treatment of cutaneous T-cell lymphoma (CTCL) where dramatic remissions in skin lesions are often observed. There have been enthusiastic claims for the benefits of photopheresis in scleroderma, although a clear indication is uncertain at this time. Photopheresis has also been used in the prophylactic management of acute graft rejection in patients with cardiac transplantation, with recent evidence of benefit.

For all apheresis procedures, it is important to ensure good vascular access. Patients should be evaluated with regard to their fluid and hemodynamic status and level of hematocrit, as there may be a substantial extracorporeal volume depending on the devices used. Vascular access is a critical aspect and one which is often neglected by the requesting physician. If vascular access by peripheral veins cannot be assured, insertion of a large lumen intravenous line into the subclavian or femoral vein is best performed as soon as possible. This is particularly important in patients who will require repeated procedures such as in TTP, myasthenia gravis, or acute Guillain-Barré syndrome, etc. Despite the removal of large volumes of fluids and the ex vivo processing, therapeutic apheresis is usually well tolerated. Side effects are often limited to hypotensive episodes, easily managed by fluid infusion of 250-500 ml saline or episodes of nausea or chills. Allergic reactions occur with the plasma infusion in TTP and are a particular problematic in patients treated with staph protein A columns. These latter patients should not be taking angiotensin converting enzyme (ACE) inhibitors, as profound hypotension linked to bradykinin has been implicated. Allergic reactions are best managed with antihistamine given intravenously.

Table 40.3. Indications for photopheresis

(a)   Cutaneous T-cell lymphoma (CTCL)
(b)   Possibly, scleroderma
(c)   Possibly, prophylaxis or treatment of allograft refection in patients postcardiac
        transplantation

Blood Transfusion in the 21st Century

Blood Transfusion practice will change early in the 21st century. Much of the focus will be on improving safety, but manufacturing products of increased and consistent potency will occur concurrently.

Universal leukoreduction of all cellular blood products (except granulocytes) and perhaps acellular products is likely to occur within the first few years. Several European countries and Canada have already mandated universal leukoreduction (1998). This is to avoid the known adverse events associated with allogeneic leukocytes, the difficulties associated with administering two inventories of blood products (leukoreduced and nonleukoreduced) and the theoretical risk of new variant Creutzfelt-Jakob disease transmission by blood transfusion (Chapter 35). This will cause the rate of nonhemolytic febrile transfusion reactions to decrease; the transmission of CMV, other herpes viruses and HTLV-1 to be reduced or eliminated, and probably a reduction in allergic reactions. The effect on reducing postoperative morbidity in surgical patients is anticipated to be an important advantage, (Chapter 13).

41

Apheresis technology will be more widely used to collect blood donations and could replace the standard manual blood collection within the first two decades in economically developed countries. This is because as the population ages, fewer donors are available and older subjects are higher consumers of blood products (Chapter 4). Maximizing the collection from each donation will be of paramount importance. When this occurs, "units" of blood will not be an appropriate request, as red cell collections will contain 380-400 ml red cells and all platelets would be apheresis derived (single donor); plasma would be in "units" of 500-600 ml, emphasizing the need to prescribe in ml/Kg (Chapter 29).

Microbial attenuation (destroying viruses and bacteria) is already advanced by 1999 and will likely make further strikes by 2010. First, acellular products (mostly plasma) is already available which is risk-reduced for viruses. This is currently achieved by the addition of methylene blue to single units with subsequent exposure to fluorescent light, the quarantining of single units, or pasteurization or solvent detergent treatment of plasma pools.

Microbial attenuation of cellular blood products is more challenging. For red cells, phtalocyanines (when exposed to red light) and psoralen derivatives inactivate bacteria and viruses. Some damage to the red cell membranes occurs, however. For platelets, psoralen derivatives show most promise. Cellular products inactivated using these processes (or others) will continue in clinical trials early in the first decade. This will greatly improve the safety of these blood products.

Recombinant plasma proteins have been available for factor VIII:C (1992) and factor IX:C (1997). Recombinant albumin and other proteins may replace plasma-derived products within a few decades.

Clinical Transfusion Medicine, by Joseph D. Sweeney and Yvonne Rizk. © 1999 Landes Bioscience

---

Table 41.1. Projected changes in blood transfusion in the 21st century

1. Universal Leukoreduction

2. Increasing Use of Apheresis Technology to Collect Blood

3. Microbial Attenuation of

    i) Acellular blood products
    ii) Cellular blood products

4. Recombinant Plasma Proteins

5. Enzymatic Conversion of Group A, B, and AB Red cells to Group O

6. Oxygen – Carrying Substitutes

7. Non Liquid (Synthetic) Platelets

Enzymatic conversion of group A and group B to group O has already been achieved experimentally and clinical studies on B → O converted red cells are ongoing. An all group O red cell supply would greatly help inventory control and improve safety by avoiding ABO incompatible hemolysis (Chapter 32).

Advances in oxygen carrying substitutes could impact greatly on blood availability for both civilian and military trauma or for patients with rare blood types, multiple alloantibodies or religious objections to blood. There has been experience with perfluorocarbon compounds (PFC) which carry oxygen in direct proportion to the amount of the emulsion in blood. PFCs may be valuable in "niche" areas (e.g., cardiac catherization with angioplasty or as a radiosensitizer for cancer cells), but more general use is unlikely. Hemoglobin-based "blood substitutes" may have a more general role. Hemoglobin is sourced from outdated human blood, animal blood or produced by recombinant technology. Hemoglobin is subjected to molecular modification by crosslinking to achieve stability, improve O_2 delivery and reduce side effects. Several such products are now in clinical studies at various stages in a variety of clinical conditions. It is likely that first generation products will be available within the next few years, initially for limited indications, particularly acute bleeding.

Nonliquid platelets are more technically difficult than synthetic red cells. Included with nonliquid platelets are cryopreserved platelets, lyophilized human platelets, freeze dried infusible platelet membranes, fibrinogen-coated albumin microcapsules and platelet glycoproteins covalently linked to the membrane of intact red blood cells. These products are earlier in clinical development than red blood cell substitutes and will likely find application for acutely bleeding thrombocytopenic patients.

APPENDIX

SOURCES OF REFERENCE

ELECTRONIC
American Association of Blood Banks web site www.aabb.org.

TRANSFUSION MEDICINE JOURNALS:
Transfusion:
> Journal of the American Association of Blood Banks
> Published in 11 issues per year by AABB, Bethesda, MD.

Vox Sanguinis:
> Journal of the International Society of Blood Transfusion
> Published in 8 issues per year by Karger, Basel (Switzerland)

Transfusion Medicine:
> Journal of the British Blood Transfusion Society
> Published Quarterly by Blackwell Scientific, Oxford

Transfusion Medicine Reviews:
> Published Quarterly by WB Saunders, Philadelphia

Transfusion Science:
> Journal of the European Society for Haemapheresis
> Published Quarterly by Elsevier Science Ltd, Devon, UK

Journal of Clinical Apheresis:
> Journal of the American Society for Apheresis
> Published Quarterly by Wiley-Liss Inc., New York

REFERENCE BOOKS:

Mollison PL, Engelfriet CP, Contreras M (eds). Blood Transfusion in Clinical Medicine, 10th Edition, Oxford: Blackwell Scientific 1997.

Petz LD, Swisher SN Kleinman S, Spence RK, Strauss RG (eds). Clinical Practice of Transfusion Medicine, 3rd Edition, New York: Churchhill-Livingstone 1996.

Speiss BD, Counts RB, Gould SA (eds). Perioperative Transfusion Medicine, 1st Edition, Baltimore: Williams & Wilkins 1997.

Rossi EC, Simon TL, Moss GS (eds) Principles of Transfusion Medicine, 2nd Edition, Baltimore: Williams & Wilkins 1996.

Index